Experimental Petrology

ALAN D. EDGAR

Experimental Petrology

Basic principles and techniques

CLARENDON PRESS · OXFORD · 1973

Oxford University Press, Ely House, London W.1

GLASGOW NEW YORK TORONTO MELBOURNE WELLINGTON
CAPE TOWN IBADAN NAIROBI DAR ES SALAAM LUSAKA ADDIS ABABA
DELHI BOMBAY CALCUTTA MADRAS KARACHI LAHORE DACCA
KUALA LUMPUR SINGAPORE HONG KONG TOKYO

ISBN 0 19 854402 2

© OXFORD UNIVERSITY PRESS 1973

PRINTED IN GREAT BRITAIN AT THE PITMAN PRESS, BATH

To Phyllis and Stephen

Preface

THE purpose of this book is to provide senior undergraduates, graduate students, and other research workers in earth sciences and related disciplines with a rudimentary knowledge of the techniques and equipment available in experimental petrology and mineralogy. Emphasis is placed on equipment and techniques rather than the implications of experimental petrology, which are discussed elsewhere. Although facilities for experimental petrology are now commonplace at many universities and other research institutions, basic information on the methods and procedures used in this type of research is so widely scattered throughout the literature in the geological, ceramic, chemical, and engineering sciences that the task of obtaining fundamental information may be extremely time-consuming. The intention of this book is to collate this information and provide a guide to the methods and procedures rather than the results and implications of the subject.

Although most of the book is devoted to methods and procedures, the initial chapter discusses the basic types of experimental equipment for various physical and chemical conditions and includes a historical review of the subject. The second chapter outlines methods of representing experimental data. The final chapter is devoted to some common problems involved in identification of the products of the experiments and determination of equilibrium reactions. It is assumed that most readers will have a knowledge of mineralogy and petrology, physical and inorganic chemistry, and elementary mathematics and physics.

The techniques described are principally those used in studies of silicate systems, and methods involving sulphide and other non-silicate systems receive only cursory attention. This omission in no way minimizes the importance of such experiments but merely reflects the fact that the author is not a specialist in this branch of experimental geology. For the same reason, techniques of determining Eh–pH diagrams, solubility data, high-temperature X-ray experiments, differential thermal analysis, etc., have been omitted or else only a brief reference is made to them.

The author is extremely grateful for the help and encouragement of a large number of people during the preparation of this book. Constructive criticism and helpful suggestions were received from the following scientists, who read various portions of the manuscript: Professors W. S. MacKenzie and W. S. Fyfe (University of Manchester), Dr. J. Nolan (Imperial College, University of London), Dr. N. D. MacRae (University of Western Ontario), Dr. A. Mottana (University of Milan), Dr. R. G. Platt (University of Edinburgh), Dr. J. M. Piotrowski (Southern Connecticut State College), and Dr. D. W. Williams (Electricity Research Council, Chester, England). Special thanks are due to Mrs. Shirley Watt and Miss Mary Douglas for typing the manuscript in draft and final form and for their constant good humour throughout this tedious task. Mrs. R. Ringsman drafted the diagrams with care and patience.

A.D.E.

London, Ontario
August 1972

Contents

1. INTRODUCTION	1
Definition and scope	1
Principles and limitations	1
Historical development	3
Types of experimental petrology	10
References	11
2. PRESENTATION OF EXPERIMENTAL DATA: PHASE RULES AND PHASE DIAGRAMS	13
Introduction	13
The phase rules	13
Gibbs phase rule	14
Goldschmidt phase rule	15
Korzhinskii phase rule	15
Phase diagrams	16
One-component systems	17
Two-component systems	20
Three-component systems	23
Multi-component systems	30
References	34
3. STARTING-MATERIALS	36
Introduction	36
Choice of starting-materials	36
Types of starting-materials	39
Preparation and testing of starting-materials	39
Glasses	39
Gels	51
Dry and wet mixtures	61
Natural starting-materials	63
Relative merits of different starting-materials	64
References	65
4. EXPERIMENTS AT ATMOSPHERIC PRESSURE	67
Introduction	67
Temperature measurement	67
Temperature scales	67
Methods of measuring temperature	68
Furnaces	76
Windings	77
Ceramic tubes	79
Quenching furnace	79
Thermal-gradient furnaces	82
Temperature controllers	83
Temperature calibration in quench furnaces	84
Sample preparation and procedure	85
Silicates	85
Sulphides	86
Experiments under inert and controlled atmosphere	87
Control of oxygen fugacity by sensing cells	89
References	93

5. EXTERNALLY HEATED PRESSURE-VESSELS — 94

Introduction — 94
Types of pressure-vessels — 94
  Morey vessels — 94
  Tuttle-type vessels — 97
Operation of Tuttle-type vessels — 100
  Pressure system and accessories — 102
  Furnaces and temperature recording and control devices — 104
  Preparation of samples and run procedures for Tuttle-type vessels — 106
Pressure-vessel performance — 110
Special uses of externally heated pressure-vessels — 115
Hazards and precautions — 115
References — 117

6. INTERNALLY HEATED PRESSURE-VESSELS — 119

Introduction — 119
Internally heated pressure-vessel designs — 120
  General principles — 120
  Yoder's (1950a) internally heated pressure-vessel — 122
  Goldsmith and Heard's (1961) internally heated pressure-vessel — 126
  Harwood Engineering Co. Inc. internally heated pressure-vessel — 129
  Internally heated pressure-vessel of Burnham, Holloway, and Davis (1969) — 133
Bridgman unsupported area seals — 134
Accessory equipment — 136
  Pressure equipment — 136
  Temperature equipment — 141
Procedures for internally heated pressure-vessels — 147
Safety precautions — 148
References — 149

7. SOLID-MEDIA APPARATUS FOR PRESSURES ABOVE 10 kbar — 151

Introduction — 151
Solid-media equipment — 152
  The Bridgman opposed anvil — 152
  The piston–cylinder apparatus — 155
  Other solid-media apparatus — 159
Calibration of temperature and pressure in solid-media apparatus — 162
  General — 162
  Pressure calibration — 163
  Temperature calibration — 165
Operation, maintenance, and safety precautions — 167
References — 168

8. CONTROL OF PARTIAL PRESSURES OF VOLATILE COMPONENTS AT HIGH TOTAL PRESSURES — 170

Introduction — 170
The buffer technique — 171
  Theory of buffers — 171
  Preparations and procedures for buffering techniques — 183
  Determination of equilibrium in buffer experiments — 186
Control of $f(O_2)$ by hydrogen–water vapour mixtures — 187
References — 189

9. PROBLEMS OF APPLYING EXPERIMENTAL RESULTS	191
Introduction	191
Identification of phases	191
Optical methods	192
X-ray diffraction methods	193
Electron microprobe	194
Miscellaneous problems of identification	195
Equilibrium	196
Determination of equilibrium	196
Equilibrium and kinetics	200
Equilibrium and starting-materials	203
Other methods of achieving equilibrium	204
Correlation of experimental results with natural data	206
References	208
APPENDIX: UNITS AND DIMENSIONS	210
AUTHOR INDEX	211
SUBJECT INDEX	214

Acknowledgements

Thanks are due to the following for permission to reproduce the material indicated, or to use it in preparing illustrative material.

Dr. F. R. Boyd (Figs. 7.2 and 7.3); Professor C. W. Burnham (Fig. 6.9); Professor H. P. Eugster (Figs. 8.1, 8.2, 8.3, 8.4, 8.5, and 8.6); Professor J. J. Fawcett (Fig. 5.4); Professor W. S. Fyfe (Fig. 9.1); Professor J. R. Goldsmith (Figs. 6.4 and 6.5); Professor H. T. Hall (Figs. 7.4 and 7.5); Professor J. R. Holloway (Fig. 6.13); Dr. E. M. Levin (Fig. 4.1); Professor W. C. Luth (Fig. 5.3); Dr. E. Roedder (Fig. 3.2); Professor R. Roy (Fig. 7.1); Dr. H. R. Shaw (Fig. 8.7); Professor O. F. Tuttle (Fig. 5.2); Dr. D. W. Williams (Fig. 5.7); Dr. H. S. Yoder Jr. (Figs. 6.2 and 6.3); American Ceramic Society (Fig. 4.1); Kent Instruments (Figs. 4.6 and 4.7); Harwood Engineering Company, Incorporated (Figs. 6.6, 6.7, 6.8, 6.10 and 6.11); McGraw-Hill Publishing Company (Fig. 7.1); Springer-Verlag Incorporated (Fig. 6.13); John Wiley and Sons Incorporated (Figs. 8.5, 8.6 and 8.7); *American Journal of Science* (Figs. 3.2, 5.2a and 6.9); *American Mineralogist* (Figs. 5.3 and 5.7); *Bulletin of the Geological Society of America* (Fig. 5.2b); *Journal of the American Ceramic Society* (Fig. 5.1); *Journal of the American Chemical Society* (Fig. 4.5); *Journal of Geology* (Figs. 6.4, 6.5 and 9.1); *Journal of Geophysical Research* (Figs. 6.2, 6.3, 7.2, 7.3, 8.3 and 8.4); *Journal of Petrology* (Figs. 8.1, 8.2 and 8.6a); *Mineralogical Magazine* (Fig. 5.4); *Review of Scientific Instruments* (Figs. 7.4 and 7.5).

1. Introduction

Definition and scope

EXPERIMENTAL petrology deals with laboratory investigations of the physico-chemical relationship of rocks and minerals under equilibrium conditions, and is generally restricted to investigations at high temperatures and pressures more pertinent to the formation of igneous and metamorphic rocks than of sedimentary ones. The aim of this book is to describe the equipment and techniques commonly used in experimental petrology, so that the graduate student or research worker unfamiliar with this branch of geology will be able to embark on a research project in this field with a minimum of supervision or literature search. It is assumed that the reader has a reasonable knowledge of mineralogy and physical chemistry as well as petrology.

Principles and limitations

The basic precept of experimental petrology is that one can reproduce in the laboratory the conditions under which rocks have formed in nature. Several assumptions must be made when the results of laboratory experiments are applied to petrological problems because of the many limitations of the laboratory methods. Some of these assumptions and limitations are discussed in the following paragraphs.

The fundamental aim of a laboratory investigation is not to *solve* petrological problems, but simply to provide further data from which the relative merits of different hypotheses for the genesis of rocks (based on field and other observations) can be evaluated. Thus, the experimental and field studies are complementary. Unfortunately there has been a certain antipathy between field and experimental workers, although each must use data collected by the other. MacKenzie (1960, p. 385) sums up the problem of this dichotomy in the following quotation: 'Just as the experimentalist must limit his use of the phrase "laboratory evidence" when considering the application of his results, so must the field geologist try to avoid the expression "field evidence", since very often this is merely his interpretation of observations which may be interpreted in a variety of ways'.

Two basic assumptions underlie the applications of the results of experimental petrology: first, that the chemistry of the rocks, under the high temperatures and pressures in which they have formed, is preserved in the rocks as they are now found at the Earth's surface; and, second, that the rocks have formed, at least locally, under conditions of physical, although not necessarily chemical, equilibrium. The general validity of these assumptions is supported by many field observations. However, these observations also point out the limitations of the experimental approach.

There are at least three inherent basic limitations in experimental petrology.

(1) *Simplicity of experimental systems compared to natural systems.* By necessity, laboratory systems are much simpler than natural systems, because it is generally impossible to design an experiment in which we can control or even duplicate the many physical and chemical variables present in nature. For example, most laboratory systems are chemically much simpler than rock systems, and only the major mineralogical components of any particular rock can be handled in the laboratory. There are two principal reasons for this: firstly, synthetic systems are based on the phase rule and, secondly, only systems of four components or less can be graphically represented, unless one phase remains saturated with water or some other volatile. These topics are discussed in Chapter 2. Another important limitation is the lack of knowledge of the chemistry of a rock at its time of formation. This is particularly pertinent for volatile components, such as H_2O, which tend to be present in the vapour phase at the time of formation of the rock, but are not preserved in the rocks. Similar physical limitations also apply to the laboratory method. Despite these limitations, the close correspondence between conditions of formation suggested from synthetic systems and those inferred from field relationships for many rocks fully justifies the experimental approach. In addition, as laboratory techniques become more highly developed, experiments are being made under conditions much more closely approaching those found in nature.

(2) *The criteria of equilibrium.* Problems of determination of equilibrium are also connected with those of the kinetics of reactions. This is an important problem in all experimental work as one of our basic assumptions is that, in nature, rocks have generally been formed under equilibrium conditions. In theory, the criteria for determination of whether a reaction represents equilibrium conditions can easily be established from thermodynamic principles. In practice, this is often difficult for a number of reasons, not the least of which is lack of basic thermodynamic data. Although values of entropies, free energies, and other thermodynamic functions necessary for determination of equilibrium are now available for many minerals at room temperature and atmospheric pressure, such functions are lacking at high temperatures and pressures. With the accumulation of accurate equilibrium data on a large number of synthetic systems, thermodynamic functions of many minerals are

gradually becoming known at elevated temperatures and pressures. When these are known there should be no excuse for not including in any experimental investigation an evaluation of the equilibrium conditions based on thermodynamic criteria. In too many published studies the diagrams of the systems are only 'synthesis' diagrams, and may bear no resemblance to the conditions of equilibrium. The problems of the criteria of equilibrium have been discussed in detail by Fyfe (1960) and are considered in the present text in Chapter 9.

(3) *Sample size and reaction times*. Closely related to problems of attaining equilibrium in laboratory experiments is the problem of the immense differences between the size of sample and the time available for reaction in the laboratory as compared with natural conditions. In the laboratory, the samples used are normally only a few milligrams, because of the necessity for accurate temperature control, the kinetics of the reaction, and the cost of sample containers. The laboratory samples are thus very much smaller than the volumes of rocks we are attempting to explain. While this may appear to be a severe limitation, it is partly compensated for by the fine grain-size of the starting-materials, which provides a large surface area for reaction and consequently increased reaction rates. Problems of preparation are considered and the most suitable starting-materials are discussed in Chapter 3.

The question of reaction rates is a difficult one and cannot be separated from the problem of determining equilibrium. Unfortunately, the formation of the majority of the Earth's crust has involved solid–solid reactions, in which the rates are much slower than solid–melt reactions. For example, the time available for a solid–solid reaction to take place in the laboratory is probably 10^6 and 10^{10} shorter than in nature. However, in many studies, no preliminary experiments are done on rate or equilibrium determinations because many apparently successful results have been obtained without these preliminary determinations. In order to determine the rate of a reaction, the mechanism of the reaction must be known, but for most reactions of geological interest such mechanisms are generally unknown. Knowledge of the reaction mechanisms may permit the wider use of catalysts, and hence reduce the seriousness of the time factor in experimental studies. Problems of the kinetics of geological reactions and examples of rate studies are given in Fyfe, Turner, and Verhoogen (1958) and Fyfe (1961).

These three limitations of experimental methods are only the most obvious and common in laboratory investigations. Provided that they are always kept in mind and every effort is made to evaluate them, they in no way detract from the tremendous importance of the experimental approach to solving geological problems.

Historical development

Although Sir James Hall (1761–1832) is generally regarded as the 'father of experimental petrology', laboratory attempts to answer petrological questions had been made a century earlier (1726) by Reaumur, who experimented on the crystallization of metals and the devitrification of natural glasses, and by Gregory Watt (1804) (son of the inventor of the steam engine, James Watt), who showed that, on cooling, heated basalt produced structures similar to the prismatic jointing found in natural basalts. Hall's experiments showed, however, the real value of the application of laboratory techniques in the solution of petrological problems. Prompted by the violent controversy between the Neptunist and Plutonist views in the late eighteenth and early nineteenth centuries, Hall undertook a series of experiments (Hall, 1805, 1812, 1826) in support of James Hutton's Plutonist view which struck a decisive blow to the Neptunist school. (Curiously, Hutton apparently rejected Hall's results.) In one set of experiments, pertinent to what we today would call igneous petrology, Hall melted fifteen samples of basalt, and by slowly cooling the fused samples showed that their textures were very similar to those of the original rock. Much later microscopic work on Hall's samples revealed assemblages of plagioclase, augite, olivine, and iron ore, typical of basalts. By far the most sophisticated of Hall's experiments were those in which he fairly successfully attempted to convert limestone into marble. By using gun-barrels as pressure-vessels, and measuring the temperatures of the heated gun-barrels with a clay pyrometer of his own design, he achieved temperatures probably as high as 1000 °C, at pressures up to 274 bars.† A detailed account of Hall's experiments is given by Eyles (1961).

Following these early investigations, experimental studies in mineralogy and petrology developed rapidly, particularly in Europe. Before the end of the nineteenth century a large number of minerals had been synthesized, and experiments carried out on a wide variety of geological phenomena ranging from the origin of ore deposits to the origin of meteorities. Much of this work was done in French, German, and Russian laboratories; A. Daubree, F. Fouque, A. Michel-Levy, J. Lemberg, J. Vogt, C. Doelter, and J. Morozewicz were the outstanding contributors to the early experimental investigations. An excellent account of this early period is given by Loewinson-Lessing (1954).

The impetus for experimental investigations during the nineteenth century was provided not only by a desire to explain geological phenomena, but also by greatly improved equipment and techniques, fostered by the industrial revolution. The most important advances in equipment

† A table of various pressure and other physical units used in this book is given in the Appendix.

in the nineteenth and early twentieth centuries were the invention of the platinum-wound electric-resistance furnace, the transformer furnace, the gas furnace, the microscope-stage resistance furnace, the thermocouple pyrometer, the optical pyrometer, and the self-registering pyrometer. Using this equipment, experiments were carried out at high temperatures and atmospheric pressure under conditions similar to those under which some extrusive rocks form.

After the discovery by Lavoisier in the 1770s that graphite and diamond were both composed of the same element, many attempts were made to produce diamond from graphite during the nineteenth century. In 1880, Hannay claimed to have produced artificial diamonds, but other scientists could not successfully reproduce his experiments and considerable controversy resulted from Hannay's work. Moissan, in 1893, also claimed to have produced diamonds as large as 0·5 mm from charcoal, but the success of these experiments is also dubious. Although the unequivocal synthesis of diamond was not accomplished until 1955 (by the General Electric Company), these early attempts provided further stimuli for experimental work, particularly the development of high-pressure techniques.

Another most important impetus for laboratory investigations of geological processes was the rapid development of physical chemistry in the latter half of the nineteenth century. By the end of this century, the analytical chemical approach to petrology was firmly established, culminating with the chemical classification of igneous rocks by Cross, Iddings, Pirsson, and Washington (1902), the compilation of analyses by Washington (1917), techniques of silicate and carbonate analysis by Hillebrand (1900, 1919), and the compilation of geochemical data by Clarke (1892, 1924). At about the same as this analytical approach to petrology was reaching its peak, petrologists were becoming much more aware of the importance of physical chemistry to petrological processes. The lack of equipment for the simultaneous application of high pressures and temperatures prevented investigation of the conditions of formation of intrusive rocks. In addition, the early work on mineral synthesis, rather than silicate systems, hindered the development of the application of physico-chemical methods to petrology until after the foundation of the Geophysical Laboratory in Washington.

The founding of the Geophysical Laboratory of the Carnegie Institution of Washington in 1907 probably marked the most important milestone in the history of experimental petrology, and shifted the emphasis from Europe to North America. The Geophysical Laboratory was successful largely because of its adequate financial resources, which were independent of political jurisdiction. More important, it was fortunate in attracting a group of distinguished scientists in the fields of physics, chemistry, microscopy, and petrology under the direction of A. L. Day. Prominent among those associated for many years with this laboratory were L. H. Adams, E. T. Allen, N. L. Bowen, C. N. Fenner, R. W. Goranson, J. W. Greig, H. E. Merwin, E. Poznjak, G. A. Rankin, E. S. Shepherd, and F. E. Wright.

With the foundation of the Geophysical Laboratory, an entirely new approach to experimental petrology was initiated. Systematic studies of melting relations of important rock-forming oxide systems were done under the most rigorously-controlled experimental conditions. New methods were developed for determining temperatures and optical properties of synthetic silicate melts and crystals, and the development of a method of quenching silicates from high temperature to room temperature permitted the study of many important binary and ternary systems. The quenching method also showed the clear distinction between congruent and incongruent† melting and solid solution in silicates, the basic processes upon which Bowen's concept of the discontinuous and continuous reaction series is based.

From these early experimental studies and from meticulous field observations, Bowen (1912, 1913, 1915, 1919) formulated the crystallization–differentiation theory for the genesis of igneous rocks, and showed the immense importance of fractional crystallization as a mechanism for their widely divergent compositions. With the publication (1928) of *The evolution of the igneous rocks*, Bowen had firmly established the importance of experimentation in igneous petrology, and had influenced petrological thought for generations to come. Perhaps of more fundamental importance was his emphasis on the necessity of the combination of field observation with laboratory methods.

Although Bowen had developed his reaction principle on the basis of experimental studies of dry systems, work had begun at the Geophysical Laboratory, prior to 1920, on designing equipment for investigation of water-bearing silicate systems under controlled conditions. In 1917, Morey and Fenner published results on the system $H_2O-K_2SiO_3-SiO_2$ at water pressures up to 80 bars which showed the importance of water in lowering the melting points of silicates, of second boiling point and critical phenomena, and of the solubility of water in silicate melts. One

---

† For definitions of these terms see Chapter 2.

of the chief problems of the early hydrothermal experiments† using externally heated pressure-vessels was failure of the vessels caused by creep and oxidation of the alloy. This was overcome by Adams (described in Smyth and Adams 1923), who designed an internally heated vessel, cooled externally by water, thus extending the pressure range of the vessels up to the cold strength of the alloys used (about 10 kbar), at temperatures almost as high as the melting-points of the furnace windings (1600 °C). Unfortunately, this apparatus was difficult to operate and maintain. Goranson (1931, 1938) used this equipment for studies of the solubility of water in granitic melts and for investigations of the $NaAlSi_3O_8$–H_2O and $KAlSi_3O_8$–H_2O systems. These studies of Morey, Fenner, and Goranson, as well as those of Greig, Merwin, and Shepherd (1933) on the transport of silica by water vapour and other volatiles, had a profound influence on such theories as gaseous transfer, hydrothermal alteration, and late-stage developments in differentiation processes, particularly in the genesis of granitic pegmatites, vesiculation, etc.

Since its founding, workers at the Geophysical Laboratory have systematically determined a large number of the simpler silicate systems, building up a firm framework for more complex systems from which the physical chemistry of natural processes is better understood. In 1927, J. F. Schairer joined the Laboratory and, in collaboration with Bowen and other workers, published results of phase studies on the systems FeO–SiO_2 (Bowen and Schairer 1932), FeO–MgO–SiO_2 (Bowen and Schairer 1935), fayalite (Fe_2SiO_4)–albite ($NaAlSi_3O_8$) (Bowen and Schairer 1936), nepheline ($NaAlSiO_4$)–kalsilite ($KAlSiO_4$)–silica (SiO_2) (Schairer and Bowen 1935), diopside ($CaMgSi_2O_6$)–leucite ($KAlSi_2O_6$)–silica (SiO_2) (Schairer and Bowen 1938), K_2O–Al_2O_3–SiO_2 (Schairer and Bowen 1947, 1955), albite ($NaAlSi_3O_8$)–orthoclase ($KAlSi_3O_8$)–anorthite ($CaAl_2Si_2O_8$) (Franco and Schairer 1951), orthoclase ($KAlSi_3O_8$)–forsterite (Mg_2SiO_4)–silica (SiO_2) (Schairer 1954), and Na_2O–Al_2O_3–SiO_2 (Schairer and Bowen 1947, 1956). Adams (1952) lists all systems studied at the Geophysical Laboratory.

Although the majority of experimental investigations in the first half of this century were concerned with magmatic processes, studies were also being carried out on systems pertinent to the formation of metamorphic rocks. By using a press of his own design, Adams (1901, 1912)

† The term *hydrothermal* is used by experimental petrologists to describe studies with a volatile component present as a liquid, vapour, dissolved liquid, or component of a solid phase. The volatile component need not be water.

investigated the flow of solid material and its implications in the processes of dynamic metamorphism. The investigations of Taylor (1934) and Taylor and Williams (1935) on the system $MgO–CaO–SiO_2$ in the absence of a liquid phase resulted in the synthesis of a number of metamorphic minerals including wollastonite, forsterite, diopside, and akermanite. In the U.S.S.R., Kurnakov and Cernych (1926) worked on dehydration reactions in chlorites and serpentine and Syromyatnikov (1934) on similar reactions in chrysotile-asbestos.

Before the end of the Second World War facilities for experimental petrology in universities were virtually non-existent, with the exception of the University of Chicago, where Bowen was the Hutchinson Distinguished Professor of Petrology from 1937 to 1945, and P. W. Bridgman's high-pressure laboratory at Harvard University. In the 1940s and 1950s this situation changed dramatically with the development by O. F. Tuttle of test-tube-type pressure vessels (early designs with hot seals, later ones with cold seals) in which high volatile pressures up to 5 kilobars could be maintained at temperatures of 750 °C for long periods of time. This equipment is described in Chapter 5, but it is worth noting here that this simple, inexpensive design, using special heat-resistant alloys, enabled a large number of petrological and mineralogical systems to be investigated under conditions much more closely approaching those in nature than had previously been possible. Its simplicity and low cost allowed its wide use by university laboratories. As Wyllie (1963, p. 94) points out 'this apparatus has provided more data of geological interest than any other type of high pressure apparatus'.

Since 1949, the development of high-pressure equipment, both for volatile and load pressures, has been extremely rapid. With the discovery of new alloys, externally heated pressure-vessels, similar in basic principle to the Tuttle vessel, can now be routinely operated at vapour pressures of up to 10 kbar and, at lower pressures, to temperatures in excess of 1000 °C. After the development of the cold-seal vessel, Yoder (1950) designed an internally heated pressure-vessel at the Geophysical Laboratory, much simpler to operate than Adams's earlier design, and capable of maintaining gas pressures greater than 10 kbar, at temperatures up to the melting-point of platinum, thus permitting experimental investigations under conditions similar to those of the lower continental crust and upper mantle of the Earth.

The post-war period also saw the rapid development of equipment for studies at very high static pressures (up to 200 kbar) at temperatures up to 2000 °C, corresponding to conditions existing in the Earth at depths up to approximately 500 km. Three basic types of equipment have been used for these 'super-pressures'. The earliest of these was the simple squeezer described by Kennedy (1955), in which the sample is subjected to pressure applied between two flat plates. In the piston–cylinder apparatus, described by Boyd and England (1960), the sample is held in a suitable container, and squeezed in a tapered cylinder by a hydraulic ram. This equipment is a modification of earlier designs by Bridgman (1946, 1949) utilizing opposing anvil-type presses to apply the pressure. Pressures up to 200 kbar, at temperatures of 1750 °C, are possible in the piston–cylinder apparatus. The third type of equipment for very high pressures is the belt-and-tetrahedral press developed by L. Coes and H. T. Hall at the General Electric Company in the 1950s, for the synthesis of diamond. With this apparatus much higher pressures are attainable than in the piston–cylinder method. Descriptions of the commoner types of high-pressure equipment are given in Chapter 7.

Using such equipment many minerals such as kyanite, sillimanite, andalusite, garnet, jadeite, and even diamond, long considered as the ultimate goal of mineral synthesis, were successfully produced. Knowledge of the conditions of formation of these high-pressure minerals has been very valuable in the petrological and geophysical interpretation of conditions in the Earth's upper mantle, where many petrological processes originate. Some of the most important processes being investigated by these high-pressure studies are the importance of orogenic pressures (in addition to hydrostatic pressures) during metamorphism of sedimentary rocks, the nature of the rocks in the upper mantle and their implications on the type of petrological transformation which takes place at the crust–mantle boundary (or Mohorovičić discontinuity), the melting of silicates under high pressures, and the geothermal gradient to be expected in the mantle. The development and importance of high-pressure research in geology has been summarized by Boyd (1964), Wyllie (1966), and Newton (1966, 1969).

In addition to the rapid progress made in the design of equipment for experimental studies in the last twenty years, new techniques have developed for controlling partial oxygen pressures, independently of total water pressure, by E. F. Osborn, A. Muan, and his co-workers at Pennsylvania State University and by H. P. Eugster and his co-workers at John Hopkins University. These techniques, discussed in Chapter 8, have proved particularly valuable in studies of stabilities of iron-bearing minerals and in investigations of petrologically important iron-bearing systems where the partial oxygen pressure controls the valency state of the iron. More recently, methods have been developed by Eugster and others to control partial pressures of other important geological volatiles, such as carbon dioxide, sulphur, fluorine, etc.

In this review no mention has been made of the historical development of experimental studies of ore-forming minerals and systems pertinent to the genesis of ores. This field has also enjoyed a rapid expansion in the past two decades under the influence of G. Kullerud and his co-workers at the Geophysical Laboratory, H. Barnes at Pennsylvania State University, P. Barton and his co-workers at the United States Geological Survey, and others. The techniques used in this type of experimental investigation are for the most part quite different from those used for silicates, and the brief mention of these techniques, given in Chapter 4 of this book, in no way belies their importance.

Many types of auxiliary equipment used in experimental studies have also been developed, particularly in the fields of microscopy and X-ray crystallography. Two outstanding examples are the high-resolution X-ray powder diffraction apparatus, using counter recording techniques, which allow more rapid identification than film methods, and the electron microprobe, permitting determination of the compositions of synthetic crystals which are only a few microns in size.

Types of experimental petrology

The historical development of experimental petrology, as outlined in the previous section, has resulted in modern laboratory studies being divided into several different types, each involving different techniques and apparatus, according to the geological environment which we are attempting to reproduce. In a book of this size it is impossible to describe the principles and techniques involved in all of these methods. Therefore, only those methods involving apparatus which is in common use are covered.

In this book, the various types of experimental petrology are subdivided according to the type of equipment required which, in turn, is related to the conditions of formation of the rocks in nature. Other methods of classifying experimental petrology have been based on the type of reaction being investigated (cf. Wyllie 1966). The four general categories of experiments are as follows.

(1) *Experiments at atmospheric pressure.* This type of investigation is pertinent to the genesis of volcanic rocks formed, at low pressures, near the Earth's surface, where any volatiles present are rapidly lost. It comprises

determination of liquidus and subliquidus† relations by the quenching method, using synthetic systems or natural rocks.

(2) *Experiments at moderate pressures (up to 10 kbar)*. These investigations are pertinent to the genesis of plutonic and metamorphic rocks except those of the highest grades of regional metamorphism. Pressures may be static or volatile (hydrothermal experiments) and cover the generally accepted pressure range of the Earth's crust and upper parts of the upper mantle. These experiments are important in an understanding of crystallization and differentiation of basaltic magma, formation of migmatites, dehydration and decarbonation during metamorphism, geothermal gradients, and many other problems. Apparatus and techniques used in this type of experiment are described in Chapters 5–8.

(3) *Experiments under high pressures (in excess of 10 kbar)*. These studies using both synthetic systems and natural material are pertinent to the geological conditions encountered in the mantle and transition zones of the Earth; they involve both static and volatile pressures. As such, they provide an aid to our understanding of the genesis of basalts, the nature of the crust–mantle boundary, the formation of very high-grade metamorphic rocks, geothermal gradients, etc. The equipment and techniques are described in Chapter 7.

(4) *Miscellaneous experiments*. This category includes experiments involving the physical properties of rocks and minerals, such as their compressibilities, solubilities of volatiles in liquids, formation of ores, and so on. In these experiments many different techniques are used, some of which are briefly referred to throughout this book.

In addition to the chapters devoted to description of apparatus and techniques (Chapters 3–8), Chapter 2 presents a very brief discussion of phase diagrams and the principles of representing experimental data, and Chapter 9 discusses identification of phrases, determination of equilibrium, and other common problems of experimental work.

References

ADAMS, F. D. (1901). *Phil. Trans. R. Soc.*, A **195**, 363.
—— (1912). *J. Geol.* **20**, 97.
ADAMS, L. H. (1952). *Am. J. Sci. Bowen Vol.* 1.
BOWEN, N. L. (1912). *Am. J. Sci.* **33**, 551.
—— (1913). *Am. J. Sci.* **35**, 577.
—— (1915). *Am. J. Sci.* **39**, 175.
—— (1919). *J. Geol.* **27**, 293.
—— (1928). *The evolution of the igneous rocks*. Princeton University Press, Princeton, N.J.
—— and SCHAIRER, J. F. (1932). *Am. J. Sci.* **24**, 177.
—— —— (1935). *Am. J. Sci.* **29**, 151
—— —— (1935). *Proc. natn. Acad. Sci. U.S.A.* **22**, 345.

† See Chapter 2 for a definition of these terms.

BOYD, F. R. (1964). *Science* **145**, 13.
—— and ENGLAND, J. L. (1960). *J. geophys. Res.* **65**, 741.
BRIDGMAN, P. W. (1946). *Rev. mod. Phys.* **18**, 1.
—— (1949). *The physics of high pressure.* Bell, London.
CLARKE, F. W. (1892). *Bull phil. Soc. Wash.* **11**, 131.
—— (1924). *U.S. geol. Surv. Bull.* no. 770 (5th ed.).
CROSS, W., IDDINGS, J. P., PIRSSON, L. V., and WASHINGTON, H. S. *J. Geol.*, **10**, 555.
EYLES, V. A. (1961). *Endeavour* **20**, 210.
FRANCO, R. R., and SCHAIRER, J. F. (1951). *J. Geol.* **59**, 259.
FYFE, W. S. (1960). *J. Geol.* **68**, 553.
—— (1961). in *The earth sciences. Problems and Progress in current research.* (ed. T. W. Donnelly) Rice University, Semicentennial publications, 59.
—— TURNER, F. J., and VERHOOGEN, J. (1958). *Geol. Soc. Am. Mem.* **73**, p. 259.
GORANSON, R. W. (1931). *Am. J. Sci.* **22**, 481.
—— (1938). *Am. J. Sci.* A **35**, 71.
GREIG, J. W., MERWIN, H. E., and SHEPHERD, E. S. (1933). *Am. J. Sci.* Ser. 5 **25**, 61.
HALL, J. (1805). *Trans. R. Soc. Edinburgh* **5**, 43.
—— (1812). *Trans. R. Soc. Edinburgh* **6**, 71.
—— (1826). *Trans. R. Soc. Edinburgh* **10**, 314.
HILLEBRAND, W. F. (1900). *U.S. Geol. Surv. Bull.* no. 422.
—— (1919). *U.S. geol. Surv. Bull.* no. 700.
KENNEDY, G. C. (1955). *Bull. geol. Soc. Am.* **66**, 1584.
KURNAKOV, N. S., and CERNYCH, V. V. (1926). *Mem. Soc. russe. Min.* Ser 2 **55**, 183.
LOEWINSON–LESSING, F. Y. (1954). *A historical survey of petrology.* Oliver and Boyd, Edinburgh.
MACKENZIE, W. S. (1960). *Liverpool and Manchester geol. J.* **2**, 369.
MOREY, G. W. and FENNER, C. N. (1917). *J. Am. chem. Soc.* **39**, 1173.
NEWTON, R. C. (1966). in *Advances in high pressure research*, (ed. R. S. Bradley) Vol. 3. Academic Press, London. p. 195.
—— (1969). *Nature* **224**, 314.
SCHAIRER, J. F. (1954). *J. Am. ceram. Soc.* **37**, 501.
—— and BOWEN, N. L. (1935). *Trans. Am. geophys. Union* (16th ann. meeting). 325.
—— —— (1938). *Am. J. Sci.* A **35**, 289.
—— —— (1947). *Am. J. Sci.* **245**, 193.
—— —— (1955). *Am. J. Sci.* **253**, 681.
—— —— (1956). *Am. J. Sci.* **254**, 129.
SMYTH, F. H. and ADAMS, L. H. (1923). *J. Am. ceram. Soc.* **45**, 1167.
SYROMYATNIKOV, F. V. (1934). *Bull. Soc. Nat., Moscow,* **42**, (Geol. Sec., Vol. 12), 137.
TAYLOR, N. W. (1934). *J. Am. ceram. Soc.* **17**, 155.
—— and WILLIAMS, F. J. (1935). *Bull. geol. Soc. Am.* **46**, 1121.
WASHINGTON, H. S. (1917). *U.S. geol. Surv., Prof. Paper* 99.
WATT, G. (1804). *Phil Trans. R. Soc.* **94**, (part 2), 279.
WYLLIE, P. J. (1963). in *High pressure physics and chemistry*, (ed. R. S. Bradley) Vol. 2. Academic Press, London. p. 2.
—— (1966). *J. geol. Educ.* **14**, 93.
YODER, H. S. (1950). *Trans. Am. geophys. Union* **31**, 827.

2. Presentation of experimental data: phase rules and phase diagrams

Introduction

THE results of experimental petrology are graphically presented in the form of phase diagrams of various types. Such diagrams are based on the Gibbs phase rule, or on modifications of this rule, and can be derived from classical thermodynamics. A discussion of thermodynamics is beyond the scope of this book, and it is assumed that readers have an elementary knowledge of this subject. The laws of thermodynamics are treated in a large number of textbooks (e.g. Glasstone 1946); other authors have concentrated on the relationship of thermodynamics to the phase rule (e.g. Ricci 1951, Finlay 1951). The applications of thermodynamics to geochemical and petrological problems are considered by Fyfe, Turner, and Verhoogen (1958), Turner and Verhoogen (1960), Barth (1962), Smith (1963), and Kern and Weisbrod (1967) among others. Levin, Robbins, and McMurdie (1964) give a useful review of the construction and use of phase diagrams as well as a glossary of the common terminology. A book on the application of phase diagrams to petrology is currently being prepared by D. L. Hamilton.

Graphical representation of experimental data becomes increasingly difficult as systems become more complex. It is not feasible to use graphical methods to depict a system with more than three components with even a single degree of freedom. In such cases, schematic methods of representing experimental data or 'perspective' drawings must be employed. Recently, computer techniques have been used to predict phase relations, in multi-component systems, from a knowledge of the thermodynamic properties of each component.

This chapter includes a brief discussion of the phase rules, and the methods of presenting and interpreting experimental data in the form of phase diagrams for one-, two-, three-, and multi-component systems. Such knowledge is just as important to the experimentalist as the methods of acquisition of his results.

The phase rules

A phase rule relates the number of phases, number of components, and degree of freedom of a system, and may allow one to predict its properties. The properties which may be predicted vary depending on the nature of the system, and consequently there are a number of phase rules, each applicable to a different situation. In many cases, equilibrium in a system involves pressure, temperature, and the concentration of the various constituents in each phase of the system; in other cases, the effects of gravitation, etc., may be important. In most experimental studies, only the variables of pressure, temperature, and concentration are considered. The Gibbs phase rule, the Goldschmidt (or mineralogical) phase rule, and the Korzhinskii phase rule are the most useful in experimental petrology. The Gibbs phase rule is fundamental to the construction and interpretation of phase diagrams involving both liquid–solid and solid–solid equilibria. The Goldschmidt and Korzhinskii phase rules are adaptations of the Gibbs phase rule for mineralogical and petrological systems.

Gibbs phase rule

The Gibbs phase rule is expressed as

$$P + F = C + 2, \qquad (2.1)$$

where P is the number of phases present at equilibrium, F is the degrees of freedom (or variances) of the system, and C is the number of components of the system. Several limitations of the phase rule should be noted.

(1) The phase rule applies only to a system at equilibrium involving homogeneous equilibrium within each phase, and heterogeneous equilibrium in coexisting phases.
(2) The phase rule does not indicate the type of components or amounts of the phases present in the system, but only their numbers.
(3) The phase rule cannot predict the kinetics (rates) of a reaction.
(4) If a system does not obey the phase rule, it cannot be at equilibrium. However, a system which obeys the phase rule is not necessarily at equilibrium, as non-equilibrium conditions may exist.
(5) Knowledge of the number of components in a system fixes the number of phases and degrees of freedom as $C + 2$. Alternatively, because the degrees of freedom cannot be less than zero, the value of $C + 2$ tells us the maximum number of possible phases coexisting at equilibrium.

For most silicate systems the vapour pressures of the solid and liquid phases are insignificant, and the partial pressures of these phases are eliminated. Similarly, graphical representation of systems with more than one component is normally shown under constant temperature (isothermal) or constant pressure (isobaric) conditions. In both cases the system has lost one degree of freedom, and the Gibbs phase rule is modified to

$$P + F = C + 1, \qquad (2.2)$$

which is known as the condensed phase rule.

The Goldschmidt (or mineralogical) phase rule

The Gibbs phase rule (eqn (2.1)) shows that the maximum number of phases in a system at equilibrium occurs only at an invariant point corresponding to a fixed temperature and pressure. Minerals and rocks usually form, however, under a range of temperature and pressure conditions corresponding to a system with two degress of freedom. Hence eqn (2.1) becomes

$$P \leqslant C. \qquad (2.3)$$

This is known as the Goldschmidt (or mineralogical) phase rule.

The compositions of most rocks can be represented by twelve oxides (SiO_2, Al_2O_3, CaO, MgO, FeO, MnO, Fe_2O_3, K_2O, Na_2O, TiO_2, P_2O_5, and H_2O). If these are considered as components, eqn (2.3) indicates that twelve, or fewer, phases may coexist in rocks under equilibrium conditions. Rocks containing this number of solid phases (minerals) are rare, because most rock-forming minerals consist of solid solutions. Therefore, the number of components, and hence the number of phases, is decreased.

Korzhinskii phase rule

The Gibbs and Goldschmidt phase rules are applicable (in the form given) only to closed systems in which components are neither added nor lost from the system. In the formation of minerals and rocks, this situation is probably rare; indeed, some components, such as H_2O and CO_2, are probably always 'mobile' in that they may enter and leave the system throughout the course of reaction. The Korzhinskii phase rule (Korzhinskii 1957) state in an open system of n mobile components, the number of minerals (phases) does not exceed the total number of components less the number of mobile components (n), or

$$P < C - n. \qquad (2.4)$$

Thompson (1955) suggests that when a rock containing a given number of 'fixed' components is open to a mobile component (n), the lowest free energy (or most stable state of the rocks) depends on pressure, temperature, and free energy of the mobile component.

$$G' = G + \Sigma\, nF_m, \qquad (2.5)$$

where G' is the lowest free energy of the rock. G is the Gibbs free energy, n is the number of moles of mobile component, and F_m is the partial free energy of the mobile component.

Phase diagrams

Phase diagrams are graphical representations of the phase rule; they show the stability of phases in a system under various conditions. An example of a simple phase diagram representing a one-component system is shown in Fig. 2.1. The conditions of stability of the phases are

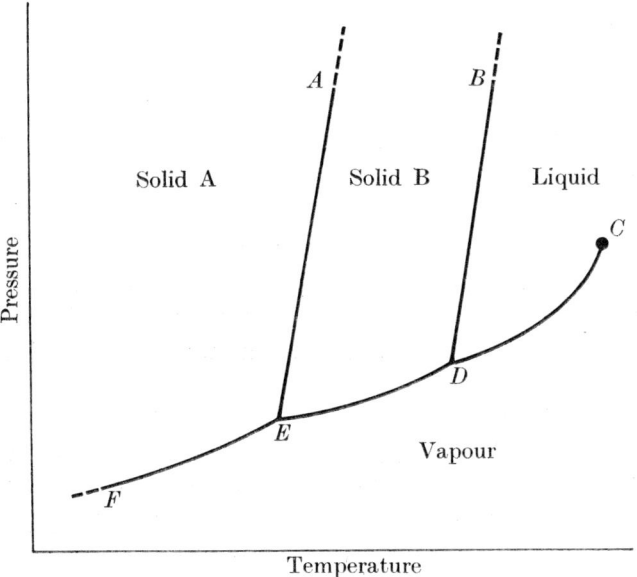

FIG. 2.1. Pressure–temperature diagram of one-component system.

plotted as functions of temperature and pressure, although any other parameters may be used. For a two-component system, graphical representation is slightly more difficult, since the relative concentrations of the two components must be considered. Such systems are commonly depicted with one of the intensive variables held constant. If pressure is constant, the diagram is then an isobaric section, and represents the conditions of stability of the phases when temperature is plotted against concentration (denoted as X), i.e. a T–X section through P–T–X space. Similarly, isothermal sections (T constant, or a P–X section), or isoplethal sections (X constant, or a P–T section) can be drawn. As the number of components increases, geometrical representation becomes more complex, because of the increased number of phases and degrees of freedom of the system. Generally, systems with more than three components cannot be represented geometrically, and mathematical methods or schematic representations must be used.

For experiments involving igneous processes, diagrams showing liquid–solid relations are most useful; for metamorphic processes, diagrams showing solid–solid relations are most commonly used.

One-component systems

The one-component (or unary) system shown in Fig. 2.1 represents the equilibrium conditions of two polymorphs, a liquid, and a vapour phase. Application of the Gibbs phase rule (eqn (2.1)) to this system shows that, in all areas in which only one phase is present, the system has two degrees of freedom (i.e. is bivariant). Along the lines FE, ED, DC, AE, and BD, two phases are in equilibrium, and the system has only one degree of freedom (i.e. is univariant). At points E and D, three phases are in equilibrium, and the system has no degree of freedom (i.e. is invariant). These are termed *triple points*.

Along FE and ED, solids A and B respectively are in equilibrium with vapour; such curves are called *sublimation curves*. BD, along which solid B is in equilibrium with a liquid of its own composition, is known as a *liquidus curve*. Along this curve, solid B melts to a liquid of its own composition; BD is therefore also termed a *congruent melting curve*. Some solids do not melt to liquids of their own composition. This phenomenon, known as *incongruent melting*, will be discussed below. DC, along which liquid is in equilibrium with vapour, is known as a *vapour pressure curve* (or *vapourus*). At some point C, along this curve, the liquid and vapour phases cannot be distinguished, and the system passes from a heterogeneous state to a homogeneous phase known as a super-critical fluid. This point is called the *critical point*, and its temperature and pressure are known as the *critical temperature* and *critical pressure*, respectively.

In one-component systems, the most stable phase may be predicted

18   *Phase rules and phase diagrams*

thermodynamically. Fig. 2.2(a) shows a simple P–T diagram with three phases A, B, and C. If the free energy per mole ($\bar{G}$) is plotted against temperature at constant pressure (p_1) (Fig. 2.2(b)), it is seen that A has the lowest free energy between 0 and t_1, B the lowest free energy between t_1 and t_3, and C the lowest free energy at temperatures greater than t_3. The intersection of the free energy curves A and B, B and C, and A and C

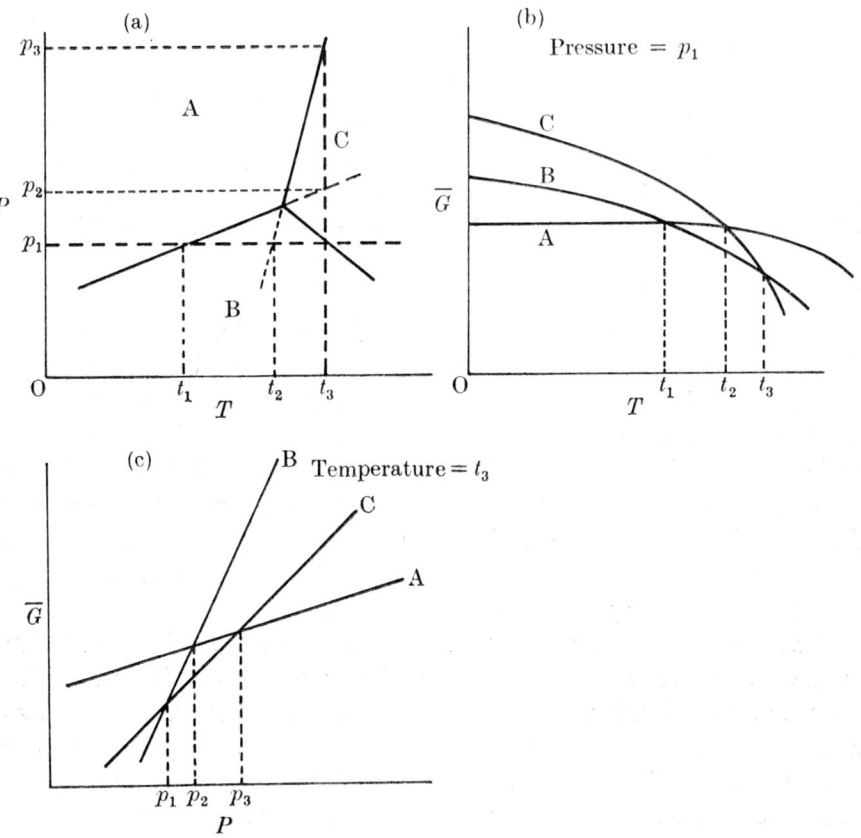

FIG. 2.2(a) Pressure–temperature diagram of one-component system. Full lines are stable boundaries between phases. Dashed extensions of full lines are metastable boundaries. (b) Gibbs free energy ($\bar{G}$)–temperature diagram for constant-pressure section (p_1) in Fig. 2.2(a). (c) Gibbs free energy ($\bar{G}$)–temperature diagram for constant-temperature section (t_3) in Fig. 2.2(a).

in Fig. 2.2(b) corresponds to the temperature of transition (metastable in the case of A to C) of these phases in the P–T diagram of Fig. 2.2(a). At the transition point, the phases are in equilibrium and their free energies are equal. This shows that the stable phase always has the lowest

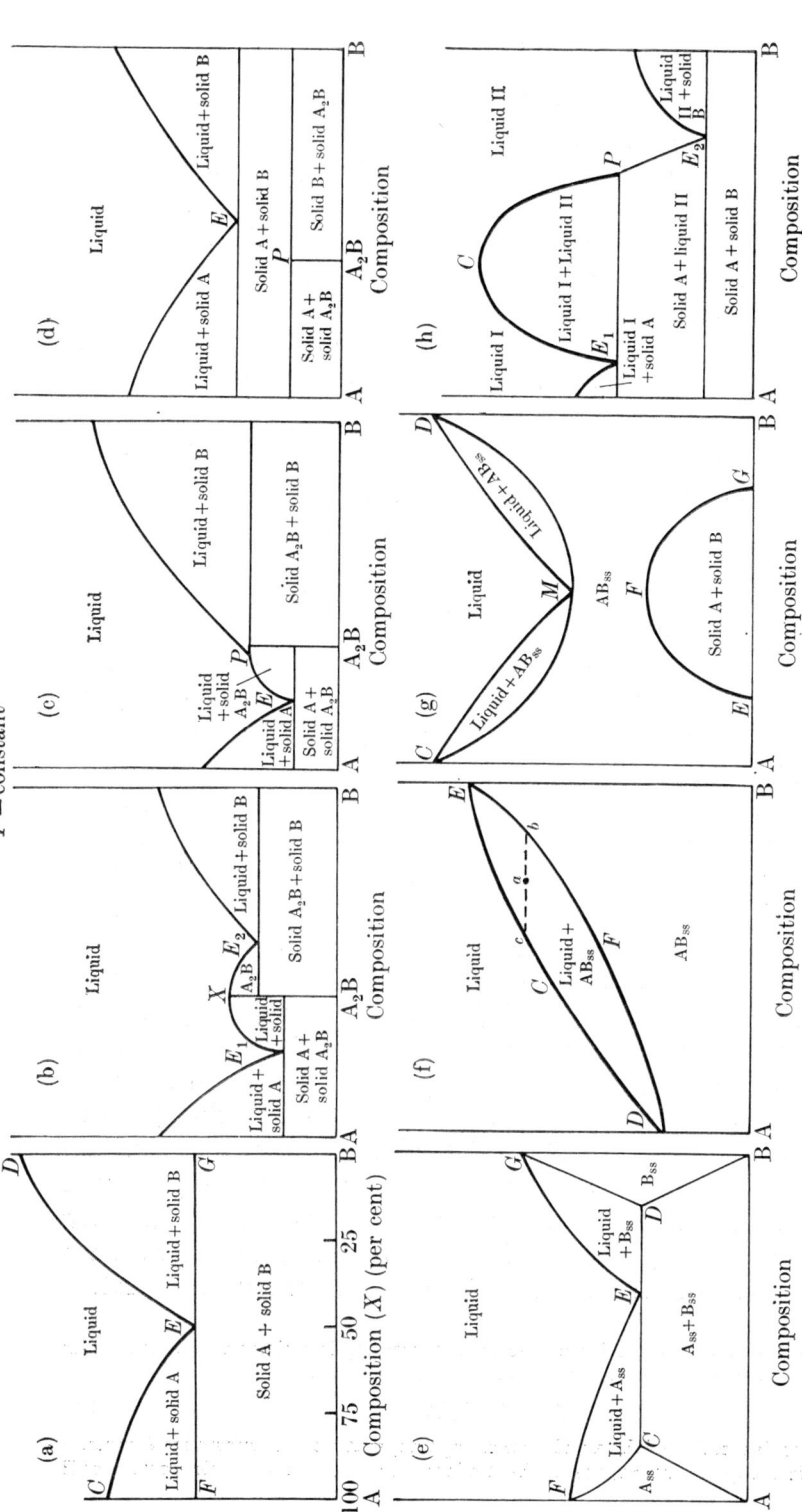

FIG. 2.3. Temperature–composition diagrams for hypothetic binary system A–B at constant pressure: (a) system with no solid solution and eutectic; (b) system with no solid solution and an intermediate congruently melting compound A_2B; (c) system with no solid solution and an intermediate incongruently melting compound A_2B which melts at temperature higher than the eutectic E; (d) system with no solid solution and an incongruently melting compound A_2B which melts at a temperature lower than the eutectic E; (e) system with limited solid solution and a eutectic; (f) system with complete solid solution and no maximum or minimum; (g) system with complete solution at high temperatures and a minimum M and solvus EFG; (h) system with liquid immiscibility in the area bounded by E_1CP.

free energy, and that the phase with the steepest free energy slope (C in Fig. 2.2(b)) is stable at the highest temperature (Fig. 2.2(a)).

A similar plot of a section at constant temperature (t_3, Fig. 2.2(a)) is shown in Fig. 2.2(c). This illustrates that the stable phase at lowest pressures has the steepest free energy curve (i.e. B in Fig. 2.2(c)). The point where the free energy curves cross in Fig. 2.2(c), corresponds to the pressure of the transitions shown in Fig. 2.2(a). From equations relating the Gibbs free energy to other thermodynamic functions, it can be concluded that high pressure favours dense phases, high temperature favours high-entropy phases; and high entropy and high molar volume correspond to high temperatures and low pressures.

Two-component systems

Representation of a two-component (binary) system is more difficult than a one-component system, because of the additional compositional variable and the formation of solid solutions and compounds. Although P–T–X relations can be shown in perspective drawings or models, petrological data are often presented as T–X sections under isobaric conditions. Several possible two-component systems are shown in Fig. 2.3. Compositions are usually plotted as weight or mole per cent along the horizontal axis and temperature along the vertical axis. The vapour phase is not represented, and its pressure is assumed to equal the total pressure, which is constant. The system is then condensed, and the condensed Gibbs phase rule (eqn (2.2)) may be applied.

In the area labelled 'liquid' only one phase is present, and the system is bivariant ($F = 2$). Points C and D (Fig. 2.3(a)) represent the melting-points of A and B respectively. With the addition of B to A, the melting-point of A is lowered as given by the liquidus curve CE,† along which liquid and solid A are in equilibrium and the system is univariant ($F = 1$). At the intersection of curves CE and DE, the minimum melting temperatures and composition are reached where two solids and a liquid are in equilibrium. This point is a eutectic, and is invariant ($F = 0$). Along FEG, liquids are in equilibrium with solids A and B, and the system is apparently invariant. However, FEG represents constant-temperature

† Provided that no solid solution is present, the lowering in melting temperature with the addition of B to A can be approximated from the equation

$$\Delta t = (RT^2/l)x, \qquad (2.6)$$

where Δt is the depression of the absolute melting temperature T of substance A when B is added, R is the gas constant, l is the heat of melting, and x is the mole fraction of B added.

conditions, and thus imposes a further restriction on the phase rule (eqn (2.2)), analogous to holding pressure constant, reducing it to $P + F = C$. This line is the solidus, and is also univariant. Fig. 2.3(a) illustrates the simplest of nine basic types of binary system, with no solid solution, no intermediate compounds, and a eutectic point. The other types are shown in Fig. 2.3(b)–(h).

Fig. 2.3(b) shows a binary system with no solid solution, but with a congruently melting intermediate compound A_2B which melts at X. This compound effectively divides the system A–B into two sub-systems, A–A_2B and A_2B–B analogous to the system shown in Fig. 2.3(a). A binary system with no solid solution and an incongruently melting compound A_2B, which does not melt directly to a liquid of its own composition but to a solid B and a liquid enriched in A at point P, is illustrated in Fig. 2.3(c). This intermediate compound decomposes at a temperature (P) higher than the eutectic temperature (E, Fig. 2.3(c)). Point P is an invariant point, termed a *peritectic* or *reaction point*, where two solid phases are in equilibrium with liquid. Fig. 2.3(d) shows a system with an intermediate compound A_2B which decomposes at P, below the eutectic temperature of the system, to solid A and solid B, giving an invariant point.

Fig 2.3(e) represents a system with a eutectic point and limited solid solution of B in A and A in B, labelled A_{ss} and B_{ss}, respectively. The limits of solid solution of both B in A and A in B increase with increasing temperature. A system showing complete solid solution without a maximum or a minimum is shown in Fig. 2.3(f), in which DCE is the liquidus and DFE the solidus. Such systems contain only one solid phase (AB_{ss}), and have no invariant point. With falling temperature, liquids along the line ECD react continuously with crystals along the line EFD, until the last of the liquid disappears at D. Fig. 2.3(g) illustrates a binary system with complete solid solution and a minimum (M) which is not an invariant point: only two phases (liquid and solid AB_{ss}) are present. In this respect the point differs from a peritectic or a eutectic point. In Fig. 2.3(g), the area below the curve EFG represents the temperature and composition at which complete solid solution cannot be maintained, and both solid A and solid B exist as separate phases with varying proportions of solid solution of one in the other. This curve is known as a *solvus*. Under certain conditions, it is possible that the solvus and solidus in such systems may intersect, producing a system with only limited solid solution and a eutectic point, as shown in Fig. 2.3(e).

A system with liquid immiscibility is illustrated in Fig. 2.3(h). In such systems, two liquids of different composition coexist in the area E_1CP. Two eutectic points (E_1 and E_2) and one peritectic point (P) are present.

There are many possible complexities in binary systems, and the ones given illustrate only the basic types. Mineralogical examples of the systems in Fig. 2 3 are given in Bowen (1928), Deer, Howie, and Zussman (1963), and elsewhere.

Crystallization courses in binary diagrams are given in Bowen (1928), Turner and Verhoogen (1960), Barth (1962), and elsewhere, and are not repeated here. The fundamental principle of interpreting phase diagrams is the Lever rule. This rule is applicable to any phase diagram, and is explained here with reference to Fig. 2.3(f). At point a liquid and solid AB coexist. The proportions of liquid to solid are found by drawing a straight line parallel to the composition axis meeting the liquidus at c and the solidus at b. From the Lever rule, the proportions of liquid to crystals are given by the ratio of the lengths of the lines ab/ac.

The formation of solid solutions in binary (and higher-component) systems may be predicted thermodynamically. In a system exhibiting complete solid solution, the change in Gibbs free energy (ΔG) is a function of temperature (T) and pressure (P) as given by

$$\Delta G = \Delta E + P\Delta V - T\Delta S, \tag{2.7}$$

where E is internal energy, V is the volume, and S is the entropy of the system. In a system of two completely miscible components A and B at

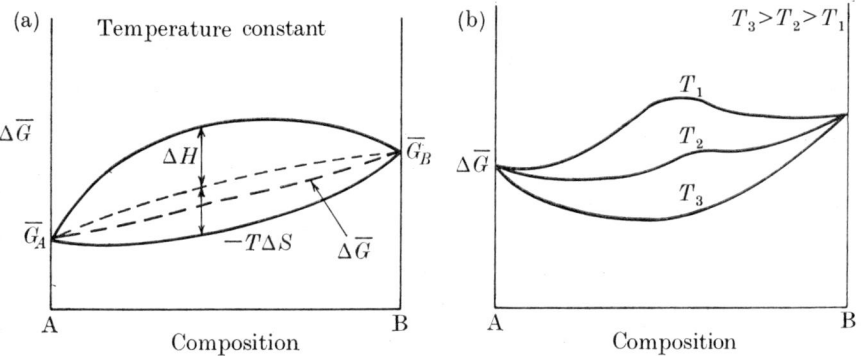

FIG. 2.4. Change in molar Gibbs free energy ($\Delta \bar{G}$) versus composition for hypothetical binary system A–B: (a) at constant temperature; (b) at various temperatures.

some constant pressure P, ΔE and ΔV will increase as compositions vary from A to B. Hence, from eqn (2.7), ΔH increases. Entropy also increases with the addition of a second component in solid solution, and the net result for $\Delta \bar{G}$ from eqn (2.7) is shown schematically in Fig. 2.4(a), where $\Delta \bar{G}$ is plotted against composition at constant temperature T. Whether $\Delta \bar{G}$ increases or decreases with changing composition depends largely on the $-T\Delta S$ term in eqn (2.7). With increasing temperature, this term becomes dominant, and a plot of $\Delta \bar{G}$ against composition for increasing temperatures (where $T_3 > T_2 > T_1$) (as shown in Fig. 2.4(b)) indicates that at high temperatures (T_3), where there is complete solid solution between A and B, the free energy curve is smooth. With decreasing temperatures (T_2 and T_1), 'kinks' develop, corresponding to decreasing solid solution. The compositional range of these 'kinks' corresponds to the position of the solvus at each temperature, such as is shown in Fig. 2.3(g).

Three-component systems

Three-component (ternary) systems have five variables, namely, temperature, pressure, and three compositional variables. However, only four of these are independent, since fixing any two compositions fixes the third. With four independent variables, representation is normally shown as an equilateral triangle (Fig. 2.5(a)), each apex of which represents the composition of the three pure components. In Fig. 2.5(a), any bulk composition, such as X, is given by the intersection of the lines ab, cd, and ef, parallel to the sides of the traingle AB, BC, and CD, respectively. Determination of compositions in a non-equilateral triangle is discussed by Levin et al. (1964).

Temperature and pressure cannot be directly represented on such a diagram. However, if pressure is kept constant, temperature can be shown on an axis perpendicular to the plane of the triangle. This is depicted in Fig. 2.5(b), which represents the simplest type of ternary system with no intermediate compounds, no solid solutions, and a eutectic point. Fig. 2.5(b) may be considered as three binary systems of the type shown in Fig. 2.3(a). A', B', and C' represent the melting temperatures of pure A, B, and C, respectively; e_1, e_2, and e_3 are the binary eutectic points of the systems, ae, b, ce_2d, and fe_3g are the solidi, and $A'e_1B'$, $B'e_2C'$, and $A'e_3C'$ are the liquidi. The addition of a third component to mixtures of the other two lowers their liquidus temperatures in this simple type of ternary system. This is analogous to the behaviour of a binary system. The lowest temperature, or ternary eutectic point, is at E, and its composition is found by perpendicular projection of E to E' in triangle ABC. Thus the areas $A'e_3Ee_1$, $B'e_1Ee_2$, and $C'e_3Ee_2$ in Fig. 2.5(b) are univariant curved ternary liquidus surfaces with A, B, and C as their respective primary phases.

24   *Phase rules and phase diagrams*

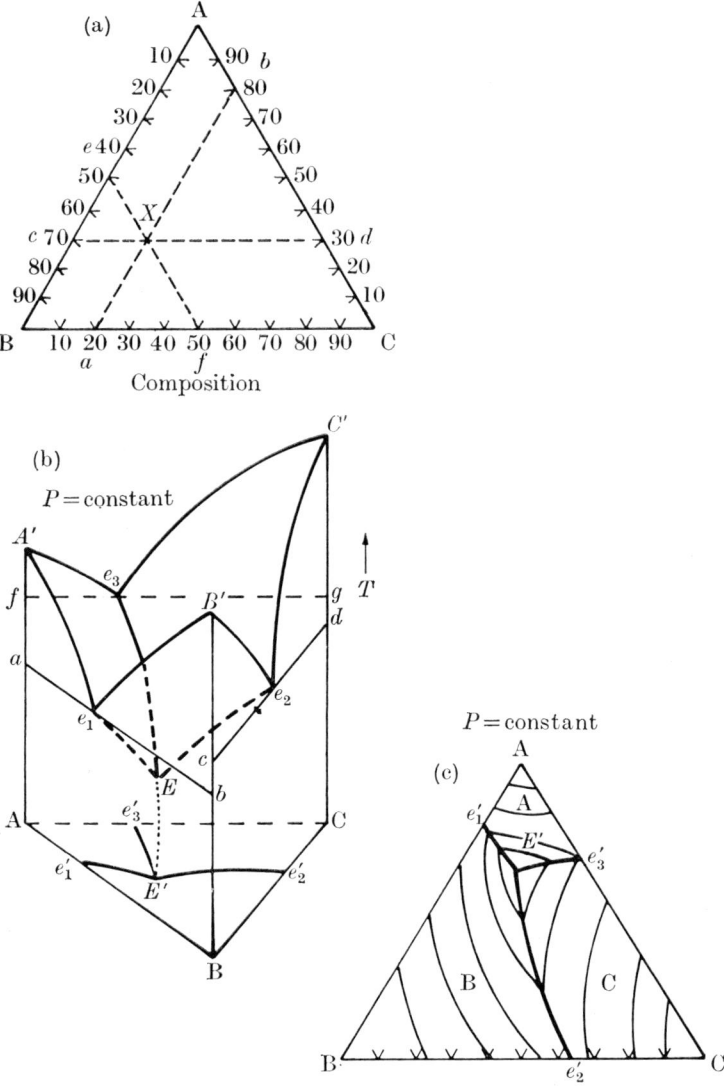

FIG. 2.5(a) Method of representing compositions of ternary systems. (b) Perspective drawing of temperature-composition at constant pressure for a simple ternary system. E, ternary eutectic; e_1, e_2, e_3, binary eutectics. E^1, projected ternary eutectic; e_1^1, e_2^2, e_3^3, projected binary eutectics. (c) Projection of ternary system shown in Fig. 2.5(b). Heavy lines are boundary curves, thin lines are isotherms.

Such 'perspective' diagrams are awkward, and commonly the liquidus (or other) surfaces are projected on to the base triangle ABC, as was described for the eutectic point E. This has been done for the triangle ABC in Fig. 2.5(b), which shows the compositional variations of the primary phases. Liquidus temperatures may also be projected and contoured, as shown in Fig. 2.5(c), which represents the projected liquidus of Fig. 2.5(b). The contours are isothermal lines, and show the shape of the liquidus surfaces. The lines $E'e_1$, $E'e_3'$, and $E'e_2'$ are boundary curves along which two solid phases are in equilibrium with a liquid. If isotherms are not given, arrows along boundary curves indicate directions of falling temperature.

Applying the condensed phase rule to the areas A, B, and C of Fig. 2.5(c) indicates that the system is divariant ($F = 2$); along the boundary curves, the system is univariant ($F = 1$); at the ternary eutectic point at E, it is invariant ($F = 0$).

Although Fig. 2.5 and the preceeding discussion have emphasized isobaric liquidus relations, similar principles apply to isothermal diagrams, or the subsolidus relations obtained by considering isothermal sections below the temperature of the ternary eutectic point.

Ternary phase diagrams can be conveniently subdivided into those with and those without solid solution. Fig. 2.5(c) represents the simplest ternary systems. Some other ternary systems without solid solution are illustrated in Figs. 2.6(a)–(e). Fig. 2.6(a) represents a ternary system with a congruently melting binary compound BC. The line *A–BC*, known as Alkemade line, effectively divides the system ABC into two ternary subsystems A–B–BC and A–BC–C each with a eutectic (E_1 and E_2). The binary system A–BC is a thermal barrier in the system A–B–C. Liquids of bulk composition to the left of A–BC finally crystallize as A, B, and BC at E_1, whereas those liquids from bulk compositions to the right of A–BC have final products of A, BC, and C at E_2. If the Alkemade line does not form a true binary system, as shown in Fig. 2.6(b), the system contains only one ternary eutectic (E) and a peritectic point (P). At P, the system is invariant and the temperature remains constant until one of the solid phases (B) completely disappears by reaction with the liquid, which may subsequently proceed toward E. Ternary eutectic points can be recognized by convergent temperature arrows on the three intersecting univariant lines, whereas peritectic points have one or two diverging temperature arrows on their univariant lines.

In Figs. 2.6(a) and (b) the binary compound BC melts congruently. An incongruent melting binary compound is shown in Fig. 2.6(c), in which

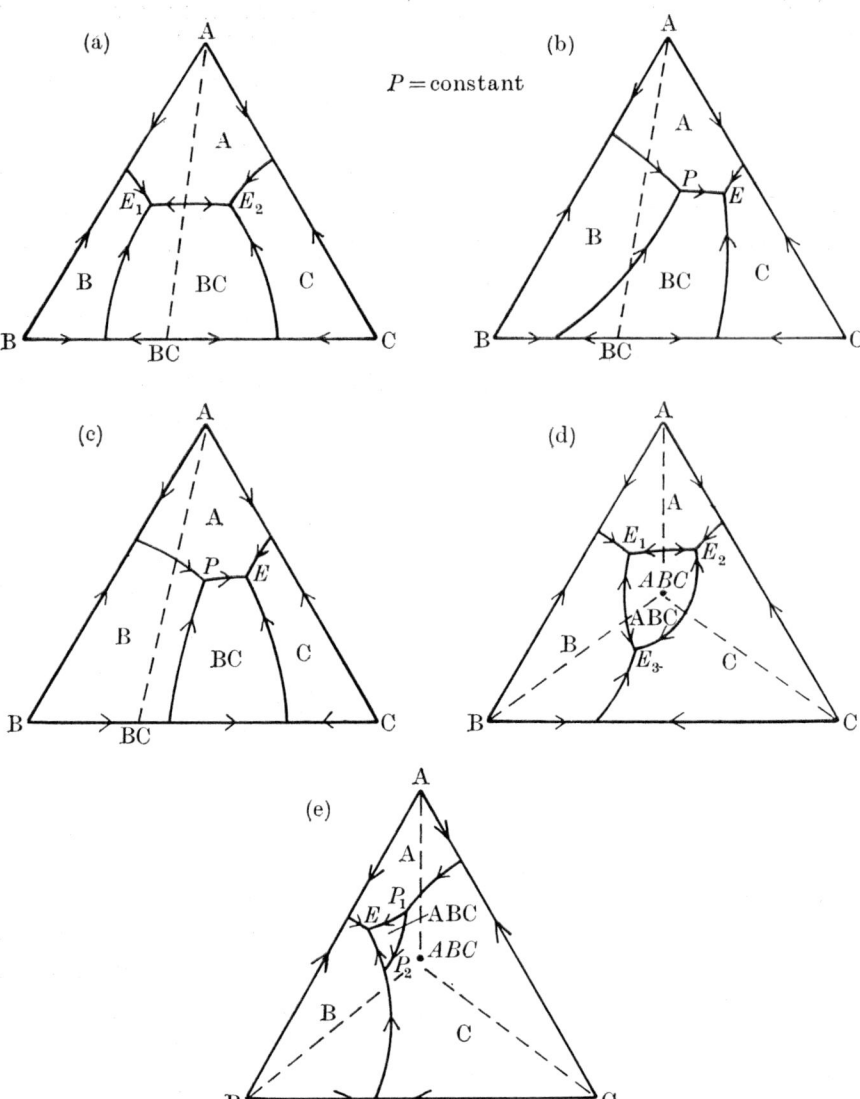

FIG. 2.6. Hypothetical ternary systems A–B–C without solid solution at constant pressure: (a) With a congruently melting binary compound BC, and eutectics E_1 and E_2; (b) with a congruently melting binary compound BC, a reaction point P, and eutectic E; (c) with an incongruently melting binary compound BC, a peritectic P, and a eutectic E; (d) with a congruently melting ternary compound ABC, and three eutectics E, E_2, E_3; (e) with an incongruently melting ternary compound ABC, two peritectics P_1 and P_2, and a eutectic E.

the primary-phase field BC does not contain the compound BC. This system contains a peritectic point (P) and a eutectic point (E). A system with a congruently melting ternary compound ABC is given in Fig. 2.6(d), which subdivides the system A–B–C into three simple ternary subsystems, ABC–A–B, ABC–B–C, and ABC–A–C. A system with an incongruently melting ternary compound ABC is shown in Fig. 2.6(e); the primary-phase field of ABC does not contain the composition ABC, and the system contains two peritectic points (P_1 and P_2) and a eutectic point (E).

Many variations of ternary systems containing no solid solution are possible. Careful application of the phase rule and a knowledge of the basic principles involved in the construction of these systems help in their interpretation.

Ternary systems with solid solutions are given in Fig. 2.7(a)–(e). The simplest type (Fig. 2.7(a)) consists of partial solid solution between one component and the other two. The limit of the solid-solution field (A_{ss}) is merely marked on the diagram. A more complex case arises with complete solid solution between two of the components of the system, as shown in Fig. 2.7(b), which may be considered as the combination of three binary systems: two, A–B and A–C, of the simple eutectic type (Fig. 2.3(a)), and a third, B–C, with complete solid solution and no maximum or minimum (Fig. 2.3(f)). Only one boundary curve (e_1–e_2), separating the primary field of A from that of BC_{ss}, occurs in this system, which has no ternary eutectic.

A system with an intermediate congruently melting binary compound BC which forms complete solid solution with A is shown in Fig. 2.7(c). The join A–BC subdivides the system A–B–C into two subsystems A–B–BC and A–BC–C of the type shown in Fig. 2.7(b). In contrast to the analogous binary system without solid solution (Fig. 2.6(a)), A–BC is not a thermal barrier, and liquids may pass from one subsystem to another. No ternary eutectic point is present in this type of system. A similar system with an intermediate incongruently melting binary compound is shown in Fig. 2.7(d); the compound BC does not lie in the primary-phase field of ABC_{ss}.

Fig. 2.7(e) illustrates a ternary system A–B–C, in which B and C form complete solid solutions with a minimum m (see also Fig. 2.3(g)). The boundary curve, separating the primary-phase fields of A from BC_{ss}, contains a ternary minimum M. To illustrate the nature of the liquidus surface of BC_{ss}, isotherms have been included for this part of the system,

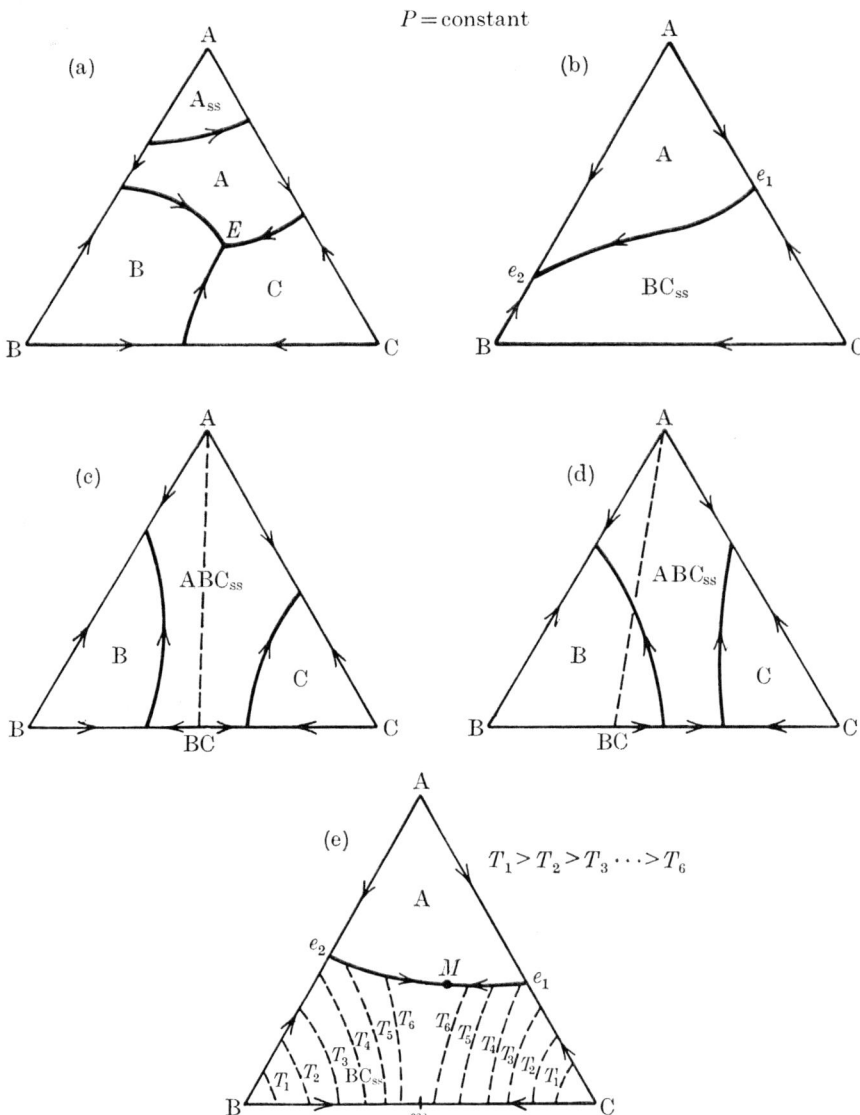

FIG. 2.7. Hypothetical ternary systems with solid solutions at constant pressure: (a) system with limited solid solution of B and C in A; (b) system with complete solid solution of B and C showing no minimum; (c) system with congruently intermediate binary compound BC showing complete solid solution with A; (d) system with incongruently melting intermediate binary compound BC showing complete solid solution with A; (e) system with complete solid solution between B and C and a minimum (M). Isotherms T_1 to T_6 indicate falling temperatures.

indicating a thermal valley running from the binary minimum m to the ternary minimum M.

Crystallization courses in ternary systems are discussed in many books and papers (e.g. Bowen 1928, Osborn and Schairer 1941, Ricci 1951, Turner and Verhoogen 1960, Barth 1962), and are not repeated here beyond considering the experimental data necessary to construct and interpret ternary systems of various types. Systems are normally determined experimentally under conditions of perfect equilibrium crystallization, and phase diagrams represent such conditions. However, courses of fractional crystallization can readily be determined from such diagrams.

In order to construct an isobaric phase diagram and interpret crystallization courses, the following experimental information is necessary.

(a) The liquidus temperatures for compositions near boundary curves and, usually, the compositions and sequence of crystallization of solid phases. From this data, the isotherms, boundary curves, and primary-phase fields are located.

(b) The temperatures and compositions of eutectics and peritectics. Often these must be determined by careful extrapolation of the boundary curves.

(c) In systems with solid solutions, the compositions of many co-existing liquids and solid phases must be determined, before crystallization courses can be worked out. On a phase diagram, the line joining a liquid in equilibrium with a solid, termed a *tie-line*, is a unique, experimentally determined line for each bulk composition.

(d) For ternary systems containing complete solid solution between two components and a minimum along the boundary curve (Fig. 2.7(e)), the position of the minimum may be determined by the method of three-phase triangles (Osborn and Schairer 1941) in which the compositions of liquids lying on the boundary curve and the compositions of the crystalline phases in equilibrium with them are joined by tie-lines forming two sides of the three-phase triangle. The third side is the line joining the two crystalline phases. The apex of the triangle lying on the boundary curve points in the direction of falling temperature. By determining a series of three-phase triangles, the minimum on the boundary may be determined.

With sufficient data, a series of isothermal sections showing the stability of various phases with temperature may be constructed, as shown in Fig. 2.8, in which the area bounded by Bngm represents liquids only, the area Amg represents liquids and solid, the area gnh represents liquids in equilibrium with solid solutions ranging in composition from h to n, the area Agh represents liquid g in equilibrium with solids A and h, and the area AhC represents solid A in equilibrium with solid solutions

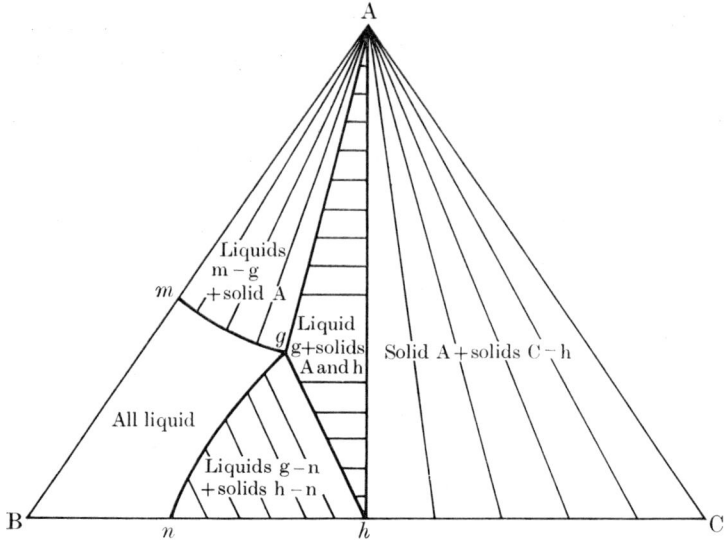

FIG. 2.8. Isothermal section through Fig. 2.7(b) at a temperature lower than that of Fig. 2.7(b). g lies on line e_1–e_2 of Fig. 2.7(b) and represents the lowest temperature at which liquids in the system can exist by themselves.

ranging in composition from C to h. A modification of this type of diagram showing stabilities of phases is the ACF and AKF diagrams used in metamorphic petrology. In such diagrams, the stabilities, and hence compatibilities, of minerals are shown.

Multi-component systems

Systems containing more than three components cannot be represented in graphical form. For example, a four-component system contains six variables: four compositional variables, of which three are independent, temperature, and pressure. Thus, if pressure is held constant and the system is condensed, four-dimensional space is required to depict the affects of temperature on the three independent compositional variables.

Simple four-component (quaternary) systems may be shown on a regular tetrahedron, each apex representing one of the pure components. Volumes within this tetrahedron represent trivariant conditions in which quaternary liquids are in equilibrium with a solid. These volumes are bounded by divariant curved surfaces, where liquids coexist with two solid phases. Three such surfaces intersect to form univariant curved lines, along which liquids are in equilibrium with three solid phases. Four univariant lines meet at a quaternary invariant point at which liquid is in equilibrium with four solid phases. In a quaternary system, an invariant point may be a eutectic, a peritectic, or a reaction point. The geometry and relationships between solid and liquid phases in multi-component systems are shown in Fig. 2.9.

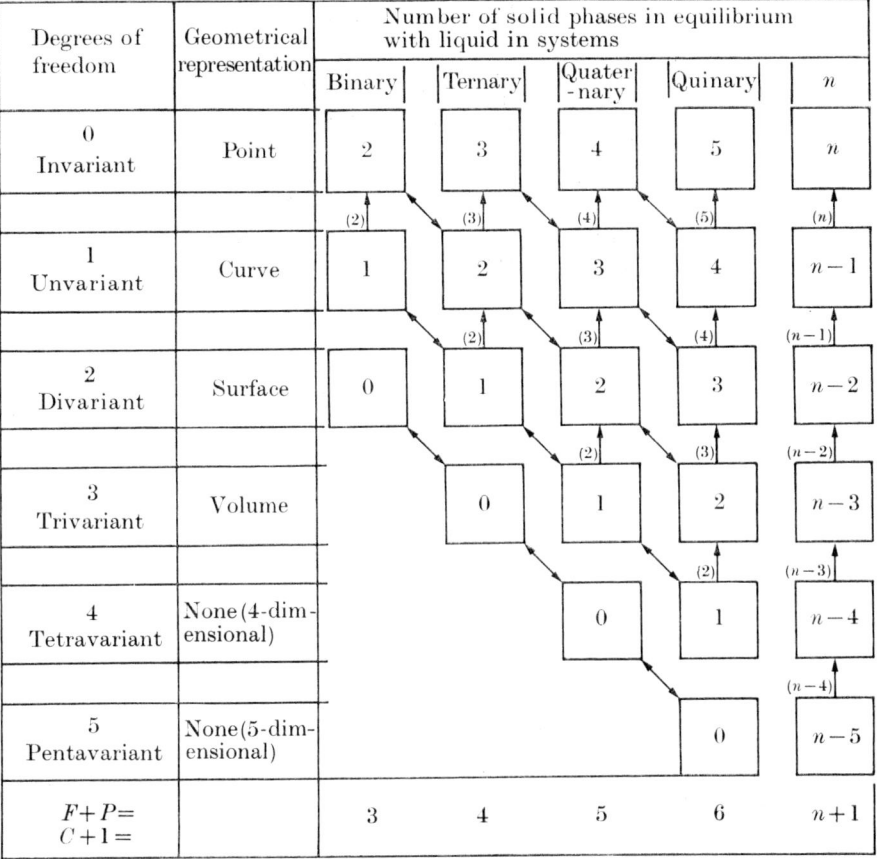

Degrees of freedom	Geometrical representation	Number of solid phases in equilibrium with liquid in systems				
		Binary	Ternary	Quaternary	Quinary	n
0 Invariant	Point	2	3	4	5	n
1 Univariant	Curve	1	2	3	4	$n-1$
2 Divariant	Surface	0	1	2	3	$n-2$
3 Trivariant	Volume		0	1	2	$n-3$
4 Tetravariant	None (4-dimensional)			0	1	$n-4$
5 Pentavariant	None (5-dimensional)				0	$n-5$
$F+P=$ $C+1=$		3	4	5	6	$n+1$

FIG. 2.9 Diagram illustrating geometrical and phase relationships in systems with up to n components. (After Levin et al. 1964.)

32   *Phase rules and phase diagrams*

In most experimental studies of multi-component systems, only the invariant points and univariant lines are determined. These are believed to be analogous to the paths taken by natural liquids forming igneous rocks. In order to locate these lines and points within the volume of a quaternary system, such as is shown in Fig. 2.10, a series of carefully

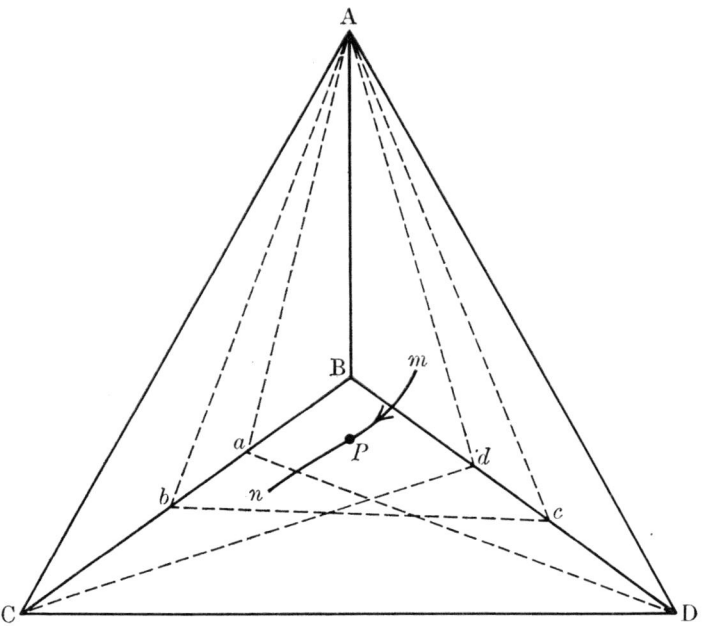

FIG. 2.10. Hypothetical quaternary system A–B–C–D. A–a–D, A–b–c and A–C–d are joins of the system; mPn is a quaternary univariant line within the volume of the tetrahedron; P is a piercing point cutting the join A–a–D.

chosen sections, or joins, through the tetrahedron are studied. In these joins, the primary-phase areas are sections through primary-phase volumes in the system, and lines separating these areas are intercepts of divariant surfaces in the system. Each point in the section where three solids and a liquid coexist represents the point at which a quaternary univariant line intersects the system. These are called *piercing points* (point P, Fig. 2.10). If the join is ternary, piercing points are eutectic or peritectic points. In contrast to true invariant points, melting or reaction will not occur sharply at piercing points but over a temperature interval, since the system has one degree of freedom.

Univariant lines in quaternary systems originate or terminate at ternary invariant points, and quaternary divariant surfaces originate or terminate at ternary univariant lines (see Fig. 2.9). From a knowledge of the four bounding ternary systems and the location of piercing points within the volume of the tetrahedron, quaternary univariant lines may thus be extrapolated. Four of these univariant lines converge at a quaternary invariant point.

Because diagrams such as Fig. 2.10 are awkward, quaternary systems are normally represented on crystallization flow-diagrams. Examples of this form of diagram can be found in many papers (e.g., Yoder and Tilley 1962, Schairer 1967, Bailey and Schairer 1966). An example of a hypothetical quaternary system and its corresponding flow-diagram is shown in Fig. 2.11. Flow-diagrams do not indicate bulk compositions

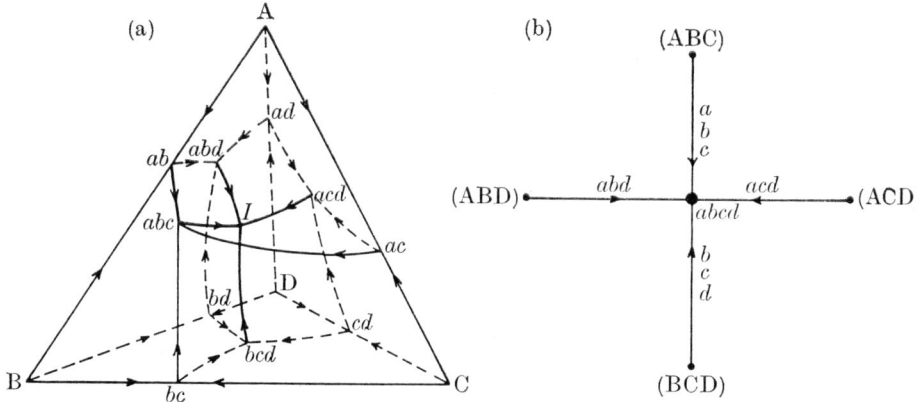

FIG. 2.11 (a) Simple hypothetical quaternary system A–B–C–D containing ternary systems A–B–D, A–B–C, B–C–D, and A–C–D with ternary eutectics at abd, abc, bcd, and acd respectively. I is a quaternary eutectic and heavy lines leading to it are quaternary univariant lines. (b) Flow diagram for system shown in Fig. 2.11(a). (Both figures after Bailey and Schairer 1966.)

along univariant lines or at invariant points within the system, but are merely schematic representations of the relationships between temperature and compositions of univariant lines and invariant points. As such, they show the crystallization trends of liquids in the system.

For systems of more than four components any type of geometrical representation is impossible and mathematical methods must be used. Although these are not considered here in detail, reference to Fig. 2.9 will indicate some of the complexities. This figure shows the interrelationships between the degree of freedom, geometry, and number of solids in equilibrium with liquids for systems containing up to n components. The numbers in the boxes in Fig. 2.9 give the number of solid phases in equilibrium with a liquid; for example, a ternary system at an invariant point has three solids and a liquid in equilibrium, whereas a ternary system at a trivariant volume has no solid phase coexisting with liquid. The vertical arrows, with numbers in parentheses, indicate the number of regions below the arrow that intersect to form the region of lower variance above the arrow; for example, in a quaternary system the intersection of two trivariant volumes produces a divariant surface. The diagonal arrows show the relationship between regions in a system and its bounding systems. Thus, a univariant curve in a ternary system may originate or terminate at an invariant point in a binary system.

Some of the relationships shown in Fig. 2.9 are obtained directly from the simple law of combination:

$$\frac{n!}{r!(n-r)!}, \qquad (2.8)$$

where $_nC_r$ is the number of combinations of r objects that can be chosen from n objects. For example, this equation can be used to calculate the number of divariant regions that form a univariant region for a system of any number of components.

Attempts have been made to determine phase relations of complex systems from thermodynamic data on their components using computer methods (e.g., Smith 1964, 1965, 1966, 1967, Blander 1968). Smith has applied these methods to mineralogical systems of three components, whose phase relations have been experimentally determined with favourable results. In theory, these techniques are applicable to systems containing any number of components.

References

BAILEY, D. K. and SCHAIRER, J. F. (1966). *J. Petrol.* **7**, 114.
BARTH, T. F. W. (1962). *Theoretical petrology*. Wiley, New York.
BLANDER, M. (1968). *Chem. Geol.* **3**, 33.
BOWEN, N. L. (1928). *The evolution of the igneous rocks*. Princeton University Press, Princeton, N.J.
CLARK, S. P. (editor) (1966). *Handbook of physical constants*, Geol. Soc. Am., New York.
DEER, W. A., HOWIE, R. H. and ZUSSMAN, J. (1963). *Rock forming minerals* (5 Vols.). Longmans, London.
FINLAY, A. (1951). *Phase rule*. Dover, New York.
FYFE, W. S., TURNER, F. J. and VERHOOGEN, J. (1958). *Geol. Soc. Am. Mem.* **73**.
GLASSTONE, S. (1946). *Elements of physical chemistry*. Van Nostrand, New York.
KERN, R. and WEISBROD, A. (1967). *Thermodynamics for geologists*. Freeman, San Francisco.

KORZHINSKII, D. S. (1957). *Physicochemical basis of the analysis of the petrogenesis of minerals.* Consultants Bureau, New York.

LEVIN, E. M., ROBBINS, C. R. and MCMURDIE, H. F. (1964). *Phase diagrams for ceramists.* American Ceramic Society, Columbus, Ohio.

OSBORN, E. F. and SCHAIRER, J. F. (1941). *Amer. J. Sci.* **239**, 715.

RICCI, J. E. (1951). *The phase rule and heterogeneous equilibria.* Van Nostrand, New York.

SCHAIRER, J. F. (1967). in *Researches in geochemistry* (ed. P. H. Abelson), Vol. 2, p. 568.

SMITH, F. G. (1963). *Physical geochemistry.* Addison-Wesley, Reading, Mass.

—— (1964). *Nature, Lond.* **204**, 370.

—— (1965). *Can. Miner.* **8**, 141.

—— (1966). *Can. Miner.* **8**, 424.

—— (1967). *Can. Miner.* **9**, 180.

THOMPSON, J. B. (1955). *Am. J. Sci.* **253**, 65.

TURNER, F. J. and VERHOOGEN, J. (1962). *Igneous and metamorphic petrology.* McGraw-Hill, New York.

YODER, H. S. and TILLEY, C. E. (1962). *J. Petrol.* **3**, 342.

3. Starting-materials

Introduction

THE preparation of samples to be used as starting-materials is of prime importance in any project in experimental petrology. The basic requirements of a starting-material are that it be of the required composition and as homogeneous as possible. If these are not fulfilled, the results of the experiments may be worthless, irrespective of the sophistication of the equipment and elegance of the techniques used. For example, in studying a three-component system, precise knowledge of the 'bulk' composition of the starting-material provides a 'reference' point in the system, exactly analogous to the coordinates of a topographical map. Physical homogeneity is important because, in most experiments, runs are made using only a few milligrams of each composition, which is normally prepared in batches each of several grams. If actual rocks and minerals are used as starting-materials, similar care should be taken in their preparation to ensure that their compositions are known and that they are as free of impurities as possible. The experimenter is therefore well advised not only to choose his starting-materials carefully, but also to take the utmost care in their preparation.

Choice of starting-materials

The choice of starting-materials depends on a number of factors, including the nature of the problem to be investigated, the availability of sufficiently pure components, the availability of equipment (such as high-temperature furnaces,† and, in some cases, personal choice. Before we describe the preparation of various starting-materials, the implications of each of these factors require brief discussion.

The nature of the problem and the experimental technique to be used are probably the most important consideration in the choice of starting-materials. In making this choice, a very basic problem arises. In purely 'synthetic' experiments (where a mineral or composition is to be synthesized, as opposed to experiments in which a natural mineral is used as

† Although furnaces are used in virtually all types of experimental work, certain starting-materials, e.g. glasses, require expensive platinum-wound furnaces for their preparation. These may not be required for the actual experiments. For further discussion see p. 41.

starting-material), it is necessary to use sufficiently reactive starting-materials, with high free energies, in order to induce the desired reaction in a reasonable length of time. Unfortunately, the most reactive materials are also those that are most likely to produce metastable phases. In theory, for the optimum possibility of achieving equilibrium, starting-materials with the lowest possible free energies should be used, but most of these materials are so strongly bonded that reactions are very sluggish. As an example of this problem, we may consider the feasibility of using natural quartz and corundum (α-Al_2O_3) as sources of SiO_2 and Al_2O_3, respectively. Both of these oxides are abundant in nature, and can be obtained in very pure form or can be easily purified. However, both quartz and corundum are in general highly unreactive and extremely difficult to dissolve, even at high temperatures, except in the presence of H_2O. If these forms of SiO_2 and Al_2O_3 are used, they must first be converted to a more reactive polymorph. The problem of the free energies of the components of starting-materials to be used in hydrothermal synthesis in sub-solidus experiments is discussed in detail by Fyfe (1960) and in Chapter 9 of this book.

In determination of liquidus relations at atmospheric pressure, in which glass is the most suitable starting-material, the problem of metastability is much less crucial than in sub-liquidus studies, because, near its melting-point, glass does not have a large excess free energy in comparison with its crystalline products, and is therefore much less likely to produce metastable phases.

In addition to the effects of metastability produced by different starting-materials, the chemistry involved in the preparation of these materials must also be considered. In order to keep the initial material 'on' composition, extreme care must be taken in its preparation, and known losses of any of its components replaced by adding further material. As an example of this problem, we may consider loss of alkalis. Alkalis tend to volatilize appreciably at temperatures greater than 900 °C, and therefore in any preparation of starting-material involving excessive or prolonged heating there is a tendency for alkalis to be lost. Although methods are available to minimize such losses, they require very careful control during each stage of preparation.

In the production of some types of starting-materials the chemistry of the reactions may not be exactly known. For example, the preparation of gels, described later in this chapter, entails precipitation reactions under carefully controlled chemical conditions.

Because starting-materials must be as pure as possible, their constituents must be carefully chosen to be of the utmost purity. Often the most obvious constituents cannot be obtained in a pure form, either because of impurities, such as heavy metals, or because they are hygroscopic. Because most methods of making starting-materials involve accurate weighing of the various constituents, it is essential that these be of stoichiometric composition. Although the manufacturers of analytical-grade reagents give analyses of impurities (such as heavy metals and insoluble residues) only rarely are accurate analyses of H_2O and other volatiles given. A much more serious problem is that of hydrated compounds in which the percentage of water often differs considerably from the theoretically expected value. All reagents should therefore be carefully dried and desiccated to prevent hydration. These problems often necessitate preparation or special manufacture of certain constituents of starting-materials, which is normally expensive and time-consuming.

In addition to the 'scientific' considerations in the choice of starting-materials, economic factors may also be important. In the preparation of high-purity starting-materials platinum containers are normally used, but for preparations at low temperatures a material such as 'Teflon' is just as suitable, and much less costly. For temperatures in excess of about 1200 °C, expensive platinum-wound resistance furnaces are normally employed in the preparation of starting materials. Such furnaces have only a limited life, and, although the platinum may be recovered and resold, the expense of equipping a laboratory with several such furnaces and other platinum-ware can be formidable. Recently, furnace windings of silicon carbide have become commercially available† which are much cheaper than platinum and can maintain similar temperatures.

In some instances, selection of starting-material may be decided by personal choice, although this is generally undesirable, for the reasons described above. Ideally, one should use as many different starting-materials of the same composition as possible, and compare the results by considering the most likely equilibrium values. Problems of the criteria for determining equilibrium are discussed in detail in Chapter 9, but it should be noted here that selecting the results from a variety of starting-materials that are in 'best' agreement with petrological observations is not a good criterion of equilibrium being achieved; as Fyfe (1960, p. 556) has pointed out, it only emphasizes the limitations of the experimental method.

† For example, 'Crusilite' and 'Crysolite', trade names of Norton International, Worcester, Mass., U.S.A.

The particle size of the material must also be considered. Although starting-materials must be sufficiently fine-grained to ensure homogeneity and to provide large surface areas for reaction to take place in a reasonable time, excessively fine-grained materials may produce disequilibrium results owing to their very large surface energies. Using glasses or gels, it is not uncommon to find unreacted starting-material in the products of the run, making location of a solidus very difficult.

The importance of choosing the correct starting-material is emphasized by Fyfe (1960, p. 556), who comments that 'perhaps some experimental geologists have been slow to realize that they can completely change the products of a synthesis by changing the type of starting-materials'.

Types of starting-materials

Four types of starting-materials are commonly used in experimental investigations: (a) glasses, (b) gels, (c) 'dry' mixtures of oxides and other compounds, and (d) natural minerals and rocks.

The first three of these may be considered as 'synthetic' types in that they are made up from pure chemical constituents to give a chosen bulk composition. For most experimental work these types of starting-materials are preferable to natural material, because a series of compositions covering a system is required, and natural materials cannot be found which are free of impurities and have the wide compositional range required.

In addition to using more than one type of starting-material, many workers employ a combination of processes in making starting-materials. For example, a gel may be fused to a glass before use, thus combining the first two processes listed above. Alternatively, one may use an analysed rock, and use the resulting glass as starting-material.

The procedures for making the basic types of starting-material containing only the common rock-forming oxides, namely, SiO_2, Al_2O_3, K_2O, Na_2O, CaO, MgO, FeO, Fe_2O_3, and TiO_2, are now given. A discussion of the relative merits of each is presented. References to starting-material containing other oxides and elements are given in Table 3.1 and 3.2.

Preparation and testing of starting-materials

Glasses

Homogeneous glasses are prepared by carefully weighing and mixing the appropriate constituents in the desired proportions and then fusing

TABLE 3.1

Sources of less-common oxides used in glass preparation

Oxide	Source	Reference to preparation
B_2O_3	B_2O_3	Greig (1927), Morey (1951), Morey and Ingerson (1937)
BaO	$BaCl_2$, $BaCO_3$	Eskola (1922), Greig (1927)
BeO	BeO	Greig (1927), Ganguli and Saha (1967)
CoO	CoO	Greig (1927)
Cs_2O	CsOH	Kracek (1930)
Li_2O	Li_2CO_3	Kracek (1930)
MnO	MnO	Greig (1927)
NiO	NiO	Greig (1927)
PbO	PbO	Greig (1927)
Rb_2O	Rb_2O_3	Kracek (1930)
SnO_2	SnO_2	Greig (1927)
SrO	SrO, $SrCO_3$	Eskola (1922), Greig (1927)
ZnO	ZnO	Greig (1927)

TABLE 3.2

Sources of less-common oxides used in gel preparation

Oxide	Source	Reference to preparation
BaO	$BaCl_2$, $Ba(OH)_2$	Ito (1968)
B_2O_3	HBO_2	Ito (1968)
BeO	$Be_2(SO_4)$	Frondel and Ito (1968)
CdO	$CdSO_4$	Ito (1968)
Cr_2O_3	$CrCl_3$	Frondel and Ito (1968)
CoO	Co	Hamilton and Henderson (1968)
Cs_2O	CsOH	Kume and Koizumi (1965)
Ca_2O_3	$CaCl_3$	Frondel and Ito (1968)
In_2O_3	$InCl_3$	Ito and Frondel (1967)
Li_2O	Li_2CO_3	Ito (1968)
MnO	MnCl	Ito (1968)
NiO	NiO	Hamilton and Henderson (1968)
P_2O_5	$NH_4(H_2PO_4)$	Ito (1968)
PbO	$Pb(NO_3)_2$	Ito (1968)
Rare earths	as oxides	Ito (1968)
Rb_2O	Rb_2CO_3	Hamilton and Henderson (1968)
Sc_2O_3	Sc_2O_3	Ito and Frondel (1968)
SrO	$Sr(NO_3)_2$	Ito and Frondel (1967)
Th_2O_3	$Th_2(NO_3)_3$	Ito (1968)
V_2O_3	Ammonium vanadate	Frondel and Ito (1968)
ZrO	Zirconyl chloride	Ito (1968)

the mixture in a platinum-crucible, usually at a temperature 50 °C to 100 °C higher than that required for complete melting. To ensure homogeneity of the glass, several fusions, with intermediate quenching and crushing, are normally required. Careful application of this procedure results in a starting-material which can easily be tested for homogeneity by refractive index measurement. Because of their tendency to undercool, silicate glasses must be crystallized partially or completely before making phase studies. If this is not done, metastable equilibrium is likely. If wholly crystalline material of the same composition as the glass is available, the presence of metastability arising from undercooling may be tested by running both materials under exactly the same conditions and comparing the results. Ideally, the crystallized glass should contain a large number of small crystals to facilitate equilibrium in a reasonable length of time.

In the preparation of glasses the required equipment depends on the number of glasses to be prepared and the size of each batch. In order to minimize weighing-errors and to reduce loss of material during preparation, glasses are normally made in 10 g batches, and should certainly not be made in quantities of less than 5 g.

The technique of making silicate glasses has been perfected by J. F. Schairer at the Geophysical Laboratory, and most of the techniques described here are given in papers by Schairer and his co-workers, particularly Schairer (1951, 1957, and 1959) and Schairer and Bowen (1955 and 1956).

Equipment required. In addition to standard laboratory equipment, glass-making requires platinum basins, crucibles, and lids, platinum foil and wire and platinum or gold tubing. The platinum crucibles should have reinforced bottoms because of the rough treatment to which they are subjected.

Two types of resistance furnaces are required in glass-making, a vertical platinum-wound furnace for making the glass and a vertical 'pot' furnace† for crystallization of the glass prior to its use as a starting-material. Suggested designs for these furnaces are shown in Fig. 3.1; their tube dimensions, etc., are given in Table 3.3. As shown in Fig. 3.1(a), the 'making' furnace consists of two tubes, the inner tube being platinum wound. The 'pot' furnace (Fig. 3.1(b)) consists of a single spiral

---

† The furnace used for making the glass may also be used for its crystallization. However, since crystallization is normally a much longer process than melting, and requires lower temperatures, it is convenient to have separate furnaces, particularly when several glasses are being made.

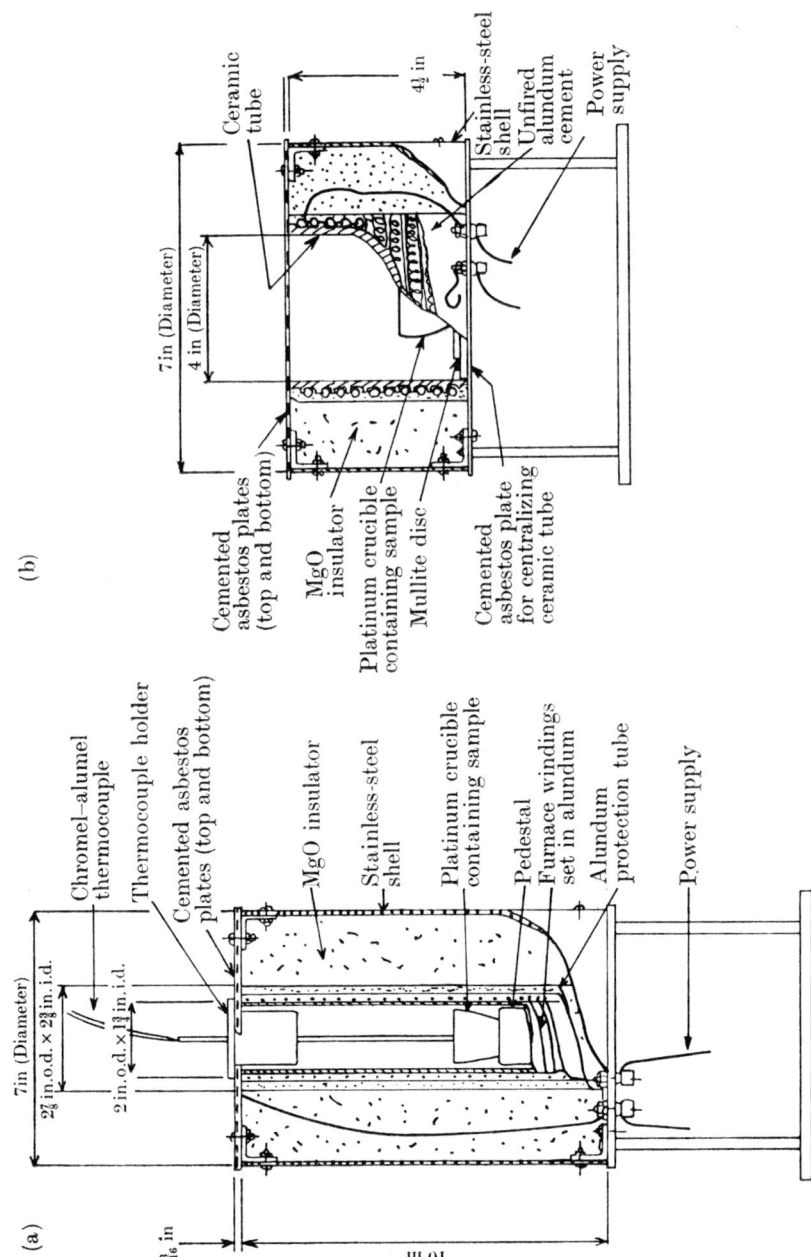

FIG. 3.1. Suggested furnace designs for glass preparation: (a) 'making' furnace for melting of premixed glass components; (b) 'pot' furnace for crystallization of glasses.

TABLE 3.3

Dimensions of furnaces for glass making†

	Platinum-wound resistance furnace		'Pot' furnace
Type	Inner tube	Outer tube	
Bore	$1\frac{3}{4}$	$2\frac{3}{4}$	4
Wall thickness	$\frac{1}{16}$	$\frac{1}{8}$	$\frac{3}{8}$
Length	12	12	$4\frac{7}{16}$
Shell diameter	9	9	14–16
Groove			
depth	—	—	$\frac{3}{16}$
width	—	—	$\frac{5}{16}$
turns per inch	—	—	$2\frac{1}{2}$‡
Wire diameter	0·031	—	0·064‡
Windings (turns per inch)	8	—	9‡

† All dimensions are in inches.
‡ Suggested dimensions for Nichrome V wire. For wires of other types dimensions will differ.

tube containing a precoiled wire. Choice of wire depends on the temperature required; platinum is used above 1200 °C; below 1200 °C other wire of suitable resistance may be used. Details of the construction of such furnaces and of temperature measurement and regulation are given in Chapter 4.

The life of a furnace depends largely on the operating temperature. If used at less than 1200 °C, a platinum wound furnace lasts several years; continuous use at 1500 °C gives a life-span of 4 to 6 months; at 1500 °C, 3 to 4 weeks; and, at 1600 °C, only a few days (Schairer 1959).

Sources of components of glasses. Recommended sources of the principal oxides of silicate glasses and their purification are as follows. (If these sources are unavailable, alternatives can be used and prepared in a similar manner.)

(1) SiO_2. Selected high-purity quartz is generally recommended as a source of SiO_2. The quartz is uniformly crushed to about 60 mesh in a steel mortar, and any iron fragments removed with a strong magnet and by handpicking. Surficial iron oxides are removed by treatment with hot dilute H_2SO_4 and hot concentrated HCl solutions, followed by washing with distilled water and drying. Determination of the SiO_2 content of the purified quartz should be made in triplicate; suitable quartz sources have average yields of 99·9 to 100·0 per cent SiO_2. The crushed and purified quartz is then placed in a platinum crucible, and roasted for approximately 2 hours at 1500 °C, to

remove any liquid inclusions. This converts it to cristobalite, which is also more readily fused with other components. It is desiccated until required.

(2) Al_2O_3. Several sources of Al_2O_3 are available. The most suitable, requiring little preparation, is T–61 Al_2O_3 made from aluminium metal by the Aluminum Company of America. Alternatively, this material may be prepared by the oxidation of pure aluminium metal, producing a product free of Fe_2O_3 and low in alkalis. Unfortunately, aluminium of sufficiently fine grain-size and purity is difficult to obtain. Aluminium 'dust' of 99·99 per cent purity has been supplied to a number of laboratories through the courtesy of the British Aluminium Company Ltd. Aluminium metal in rods and other coarse forms, even of high purity, is not recommended; it is almost impossible to reduce it to a suitable size for treatment with HNO_3 or H_2SO_4 without introducing contamination.

The fine aluminium dust is converted to $Al(NO_3)_3$ in a beaker by reaction with hot concentrated HNO_3 at about 80 °C; complete reaction of 5 g o-dust requires about 24 to 48 hours under these conditions. The resulting solution is evaporated to dryness, and washed with distilled water several timesf with intermediate evaporations. The nitrate is converted to the oxide by gently heating with a Meker burner until all nitrogen oxide fumes diappear. Finally, a 4-hour roasting at about 1400 °C converts the oxides to corundum.

If finely divided aluminium is unavailable, coarser forms may be dissolved in HCl, and subsequently the $AlCl_3.6H_2O$ may be treated with HNO_3 to convert to the nitrate. Alternatively, $AlCl_3.6H_2O$ may be converted to γ-Al_2O_3 by heating below 700 °C or to corundum (α-Al_2O_3) by strong roasting. A third source of Al_2O_3 is chemically-precipitated $Al(OH)_3$. Unfortunately, this material usually contains alkalis, which must be removed by treatment with alkali-free NH_4OH, followed by removal of the excess NH_3 and conversion to corundum by heating. The dried material usually contains less than 0·1 per cent alkalis (Schairer and Bowen 1955).

After preparation, the Al_2O_3 is stored in a desiccator until used.

(3) K_2O. Suitable sources of K_2O and Na_2O are difficult to obtain, and in glass preparation are generally added as the crystalline silicates, $K_2Si_2O_5$ and $Na_2Si_2O_5$, respectively, giving an additional source of SiO_2. These preparations are tedious, but need only be done periodically as sufficient K_2SiO_5 and Na_2SiO_5 can be prepared for several batches of glasses Details of the preparation of these compounds is given by Schairer and Bowen (1955, 1956), from which the following description is condensed.

Dry, powdered silica and $KHCO_3$, preferably of reagent grade, are weighed into a platinum crucible in the required proportions and mixed with a platinum spatula. In order to combine the $KHCO_3$ and SiO_2 to give $K_2Si_2O_5$, the crucible with lid must be carefully heated at a series of temperatures. Rapid heating is to be avoided as evolution of H_2O and CO_2 produce frothy melts, which foam out of the crucible, and also result in considerable loss of K_2O by volatilization. The recommended procedure is to heat for approximately five days in the 'pot' furnace, at a temperature slightly less than 742 °C; this is the minimum melting-temperature in the system $K_2O.SiO_2$–SiO_2 (Kracek, Bowen, and Morey 1929). At this temperature, no liquid forms, but reaction rates are fast and there is a rapid evolution of CO_2 and H_2O, but negligible loss of K_2O by volatilization. The temperature is then raised to 800 °C, 900 °C, and 1000 °C for successive 24-hour periods, producing slow melting but no violent gas evolution. Finally, the crucible is transferred to the platinum-wound furnace and placed in the predetermined 'hot spot' by means of a corundum disc pedestal. Transfer of the crucible in and out of this furnace may be facilitated by inserting a piece of platinum wire through two large pin-holes in the top of the crucible and using this wire to lift and lower the crucible. The temperature is slowly raised to 1200 °C to effect complete melting. When melting is completed, the crucible is quenched by plunging it into a shallow pan of water filled to about one half of the depth of the crucible, and immediately placing it in a desiccator with a tight-fitting lid. Because of their extremely hygroscopic nature and tendency to decrepitate, desiccation is very important in the preparation of alkali silicate glasses. A clear glass should be present at this stage.

Loss of K_2O by volatilization is now determined by weighing the crucible and contents. Schairer and Bowen (1955) found that loss of weight amounted to only 10 to 20 mg for an original 20 g sample. This loss is replaced by adding additional $KHCO_3$ in the correct amount, followed by decomposition of the bicarbonate at 750 °C for 12 hours, and by three 1-hour fusions and quenchings at 1200 °C, with intermediate crushings. Steel chips from the mortar are removed by a magnet.

Because of the extremely hygroscopic nature of the glass, it must be completely crystallized immediately after preparation, by heating the crushed glass at 750 °C for 24-hour periods, with intermediate crushings, until the powder is no longer fritted. Finally, the crystalline $K_2Si_2O_5$ is stored in a desiccator over KOH sticks, to prevent addition of CO_2 and moisture.

(4) Na_2O. The preparation of $Na_2Si_2O_5$ as a source of Na_2O and SiO_2 is similar to that of $K_2Si_2O_5$, and is described by Schairer and Bowen (1956). In this case, the required proportion of high-grade Na_2CO_3 (or very pure $NaHCO_3$ heated to 200 °C and dried) and SiO_2 is heated for two days at less than 789 °C, to remove CO_2 and H_2O. This temperature is the minimum melting-temperature in the system Na_2O–SiO_2 (Kracek 1939), and produces rapid reaction with no frothing or loss of Na_2O by volatilization. As described for $K_2Si_2O_5$ preparation, fusion is effected by heating at progressively higher temperatures up to 1200 °C, quenching the glass, and determining any loss of Na_2O by weighing the crucible. Losses, which are generally much less than with $K_2Si_2O_5$ glasses, are corrected by adding the appropriate amounts of Na_2CO_3, which is decomposed by heating at 750 °C for 12 hours. Finally three 1-hour fusions with intermediate crushings at 1200 °C give a homogeneous glass of $Na_2Si_2O_5$.

Because of the hygroscopic nature of this glass, it is immediately crystallized at 800 °C for successive 24-hour periods, and the product examined and stored in the same manner as described for $K_2Si_2O_5$.

(5) CaO. $CaCO_3$, preferably of 'spec.' pure grade, is used as a source of CaO. It must be dried for five hours at 500 °C, and kept desiccated.

(6) MgO. 'Chemically pure' oxide is used as a source of MgO. This always contains moisture and carbonate and should be ignited at 1050 °C for 2 hours to remove H_2O and CO_2, and finally at 1500 °C for 1 hour to obtain an anhydrous product, which is kept desiccated.

(7) FeO. Systems involving elements existing in more than one oxidation state are difficult to prepare owing to problems of controlling and defining their exact oxidation condition. Of these elements, iron is geologically the most important. A number of methods have been devised for experiments with iron-bearing systems and are discussed in later parts of this book, particularly in Chapters 4 and 8. Principally because of its tendency to oxidize readily, most investigators use an iron oxide in a higher oxidation state than that of FeO, with subsequent adjustments in the compositions of their glasses. For example, Roedder (1952) used reagent-grade Fe_2O_3 heated in air at 1400 °C for 10 minutes, producing Fe_3O_4 with $2\frac{1}{2}$ per cent 'excess' oxygen over that given by this formula. Presnall (1966) recommends heating reagent-grade Fe_2O_3 for 18 hours at 900 °C to produce Fe_3O_4.

A further problem in obtaining a suitable source of FeO is that iron tends to alloy with platinum at high temperatures (producing an effect exactly opposite to that of oxidation). This, however, can be avoided by heating for longer periods at lower temperatures, or by analysing the heated product and replacing any lost iron.

(8) Fe_2O_3. 'Specroscopically' pure Fe_2O_3 may be used as a source of Fe_2O_3

(9) TiO_2. Reagent grade TiO_2 is used as a source of titania.

Sources of the components of glasses containing other oxides are given in Table 3.2.

Preparation of glasses. The preparation of glasses is described in two parts: those containing no iron oxides are considered followed by those containing iron oxides. Temperatures required for complete fusion of a glass depend on its bulk composition. Generally, a temperature approximately 100 °C to 200 °C higher than the 'expected' fusion temperature will produce complete melting of refractory components such as Al_2O_3, and rapidly attain equilibrium. The 'expected' fusion temperatures are obtained by extrapolation from published liquidus diagrams of related systems, or, if necessary, by 'trial and error' methods.

(a) *Glasses without iron-oxide components.* The required proportions of each constituent of the glass are calculated, remembering that alkali-bearing compositions will have part or all of their SiO_2 content contributed by the alkali silicate. The components are then carefully weighed to at least ± 0.0002 g on a watch-glass. Because of its tendency to 'creep', the light-weight silica is weighed directly into a 25-cm^3 platinum crucible. The ingredients are then thoroughly mixed on glazed paper with a platinum spatula and transferred to the crucible, which is covered, and placed on the refractory pedestal in the 'hot spot' of the platinum-wound resistance furnace by means of a length of platinum wire as described above.

In mixtures containing $CaCO_3$, the carbonate is decomposed by initial heating at 1000 °C for 2 hours, followed by further heating at 1250 °C for 2 hours. The temperature is then raised to at least 100 °C, and possible as high as 200 °C, higher than the complete fusion-temperature of the mixture. Compositions with large amounts of Al_2O_3 and those which produce very viscous melts require high fusion-temperatures. Unfortunately, prolonged heating at high temperatures tends to produce loss of alkalis by volatilization. This may be avoided by keeping fusion-times short (1 to 3 hours) and by frequent quenching and crushing of the glass, followed by further heating. In order to dissolve the Al_2O_3 completely as many as six to ten fusions, with intermediate quenchings and crushings, may be necessary.

Rapid quenching of the glass is achieved by removing the crucible from the furnace, and immediately plunging it into a shallow dish of water to about half of the depth of the cruible. During this process, care must be taken that the lid of the crucible fits snugly to prevent loss of material by decrepitation. The contents of the crucible are removed by tapping it with a small hammer. During the early fusions, when considerable undissolved Al_2O_3 remains, the glass should be only roughly crushed in a hardened steel mortar to prevent any loss of material. At later stages of fusion, when only small amounts of undissolved Al_2O_3 remain, the glass is finely crushed to a uniform powder. After all crushings, and before refusion, any steel fragments from the mortar are removed with a strong magnet and by hand-picking. After each crushing, and before returning the crucible and its contents to the original fusion temperature, the crucible is heated to approximately 1000 °C (or some temperature below the original fusion-temperature) for about 10 minutes, to prevent material being lost by decrepitation.

During the fusion and crushing process, the homogeneity of the glass can be readily detected by placing a small amount of powdered glass on a microscope slide, adding a drop of liquid of refractive index about 0·005 higher than that of the glass, and placing a cover slip on the slide. The slide is then heated with a microscope lamp, and the presence of any inhomogeneity (such as undissolved Al_2O_3) can readily be detected with a petrographic microscope. On cooling, when the exact refractive index of the glass is reached, a multitude of tiny glass grains simultaneously disappear. Until this happens, the glass is not homogeneous and further crushings and fusions are required. This method is the best one known for determining the homogeneity of glasses. As little as 0·5 per cent undissolved Al_2O_3 can be detected, an amount about ten times less than is detectable by X-ray diffraction methods.

The magnitude of the alkali-loss by volatilization, during the frequent fusions necessary to dissolve Al_2O_3, is partly dependent on the other components of the glass. For example, the addition of CaO and MgO to an alkali–silicate melt tends to increase this loss (Madorsky 1931; Hignett and Royster 1931; Goldsmith 1949; Schairer, Yoder, and Keene 1954; Schairer and Bowen 1956). In contrast, the addition of Al_2O_3 to an alkali–silicate melt stabilizes the alkalis at high temperatures (Schairer and Bowen 1956).

Because silicate glasses tend to undercool, it is necessary to have some crystals present in the glass, before using it as a starting-material for phase-equilibrium studies. To accomplish this, portions of each glass are held in the 'pot' furnace at 50 °C to 100 °C below their 'estimated' liquidus temperature; the 'estimated' temperature is determined by extrapolation from nearby points, or by 'trial-and-error' methods. The time required for crystallization depends on the viscosity (and, hence, composition) of the glass; viscosity decreases with temperature for a given composition. For melts with low viscosity, the time required for crystallization may be only a few minutes, whereas in highly viscous melts crystallization may take months or even years. The ease of crystallization also depends on the amount of overheating necessary to dissolve Al_2O_3. Severely overheated glasses may not crystallize, even after several years treatment at temperatures well below their known liquidus. Schairer (1959) suggests 'acclimatizing' the glass at successively lower temperatures with frequent crushings, until, at a temperature 50 °C to 100 °C below the liquidus, the glass crystallizes in a few days.

In order to facilitate crystallization, it may be necessary to use 'hydrothermal' techniques (as described in Chapter 5) or, if equipment for these techniques is not available, hydrothermal crystallization can be brought about as described by Schairer (1959). In this method, the glass is placed in a platinum-foil envelope of suitable size together with a similar envelope filled with obsidian containing traces of water and other volatiles. Both envelopes are then enclosed in a fused silica tube, which is evacuated, sealed, and placed at a suitable temperature for a few days. On heating, the volatiles from the obsidian are driven into the atmosphere of the silica tube and absorbed by the glass. The addition of this mineralizer is sufficient to promote crystallization in a short period of time.

During the crystallization process, identification of phases may be made by optical and X-ray diffraction techniques. Observation of the crystallizing phases serves as a rough guide in locating primary-phase fields and estimating their temperatures for the quenching experiments.

The partially crystallized glasses should be desiccated until used.

(b) *Glasses containing iron oxide components.* As mentioned above, preparation of glasses containing iron oxides (particularly FeO) is much more difficult. The most commonly used method is an adaption of that developed by Bowen and Schairer (1932), in which the required components are fused in iron crucibles in an inert atmosphere (such as pure nitrogen). By analysing the resultant glass and correcting for iron loss, it is possible to prepare glasses of known bulk compositions. Attempts to prepare ferrous oxide-bearing glasses in a vacuum furnace have been largely unsuccessful. By using suitable gas mixtures in the furnaces used for glass preparation, it is possible to maintain the correct $f(O_2)$ to preserve the oxidation state of iron. These furnaces are described in Chapter 4.

Roedder (1952, 1965) has described a modification of Bowen and Schairer's original technique for the preparation of such glasses. The principle of this method is that by fusing mixtures in a 'neutral' atmosphere the iron may be reduced to the ferrous state. However, by using a gas such as nitrogen, there is also the danger of complete reduction of the iron oxide to the metallic state. This is compensated for by fusing the mixture in an iron crucible. A synopsis of the method described by Roedder (1952) is given below.

All components, other than 'FeO', are carefully weighed, and transferred to an agate mortar. Approximately the correct weight of Fe_3O_4 (see p. 46) is added, and all the components are thoroughly mixed. To this mixture is added about 30 per cent of the total weight of the batch of fine pure-iron chips, and the whole is transferred to a 20-cm³ pure-iron crucible with a tightly fitting lid. In order to eliminate small amounts of localized diffusion of iron from the crucible wall during fusion, the iron crucible is placed before heating in a vacuum desiccator, in which the air is replaced by pure nitrogen. The mixture is then heated at 1450 °C for $\frac{1}{2}$ to 1 hour in a pure nitrogen atmosphere, in a special furnace adopted for this purpose. Details of this furnace are given in Fig. 3.2.

The crucible and contents are quenched in mercury, the iron chips separated from the glass with a magnet, and the glass finely crushed in a Plattner-type mortar. The iron chips and glass are then returned to the crucible, and the process repeated until an optically homogeneous glass

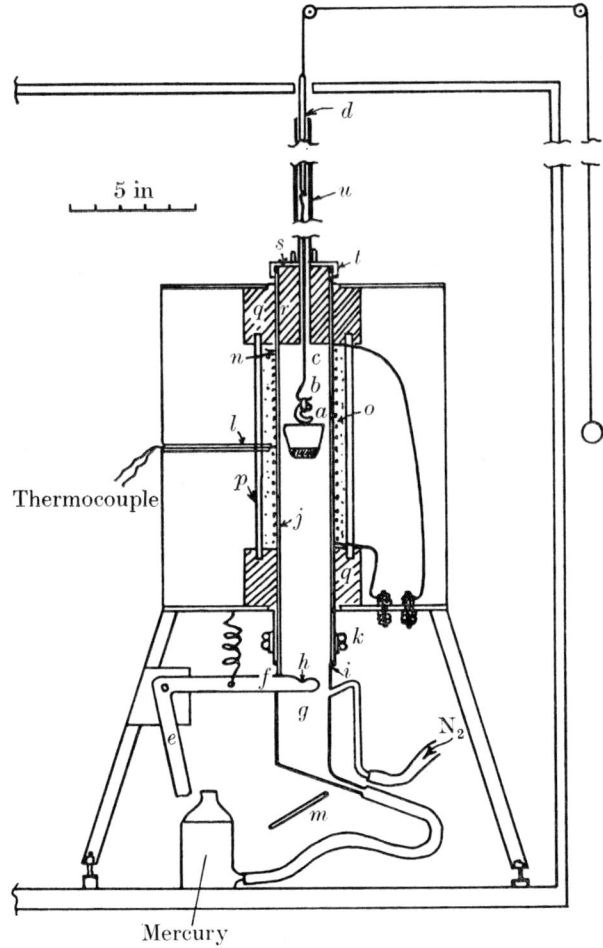

FIG. 3.2. Furnace for preparation of iron-bearing glasses. (After Roedder 1952; details of components of furnace are given in this paper.)

is obtained. For many compositions, only a few fusions are necessary, because of the excellent fluxing action of the FeO produced during initial melting. During the first fusion, part of the oxygen in the Fe_3O_4 is reduced by combination with the metallic iron, producing additional FeO ($Fe_3O_4 + Fe \rightarrow 4FeO$). In order to calculate the amount of Fe_3O_4 to be added initially the, required FeO is therefore multiplied by 0·75.

This factor is partly due to the stoichiometry of the question related to the combination of Fe_3O_4 and metallic iron, and partly due to the 'excess' of $2\frac{1}{2}$ per cent oxygen in the Fe_3O_4 (see p. 46). This factor does not account for the 'oxygen' content of the melt (as Fe_2O_3) in equilibrium with metallic iron. Each melt therefore contains variable amounts of Fe_2O_3 and not necessarily the correct FeO content.

In order to obtain the exact composition of each glass, it is necessary to analyse it for total iron and for Fe_2O_3, which may be more or less than required. For some FeO-bearing glasses the ratio of the other components is fixed (provided there is no loss during fusion, etc.); by analysing for total iron and Fe_2O_3 the bulk composition can therefore be found.

After the final fusion, any chips of metallic iron or steel chips from the mortar are removed with a magnet and the glass is analysed. The remaining glass is crystallized either in a vacuum furnace or in a nitrogen atmosphere, using iron containers. The temperature and time required for crystallization are similar to those for non-iron-bearing systems, described above. The partly crystallized glass is desiccated until required.

Determination of glass compositions. With any starting-material, the investigator is faced with the question, 'Is it "on" composition?' One method of answering this question is, of course, to analyse chemically each starting-material. However, the analytical precision is not always high enough to detect errors of significant magnitude.

Routine analyses of the constituents most likely to be lost (K_2O, Na_2O, FeO, and Fe_2O_3) are made during the manufacture of glasses, and the presence of undissolved Al_2O_3 is readily detected by the optical method described above. In addition, by determining the refractive indices of the glasses and plotting isofract curves (lines joining glasses of equal refractive index), an anomalous composition can easily be detected, for these indices vary systematically with composition. Similarly, if liquidus determinations are being made, any glass not 'on' composition will show up as an anomaly in the otherwise smooth liquidus curves. Thus, although chemical analysis can be done on each glass, physical methods are usually just as effective in detecting glasses with incorrect compositions.†

Gels

The tedious procedures required in preparing glasses, the problems of loss of alkalis at high temperature, and difficulties in nucleating

† If equipment is available, certain glass compositions, particularly those low in alkalis, may be determined by electron-microprobe analysis.

low-temperature phases from their high-temperature structures lead to methods of making starting-materials as gels.† This method, originally developed by Barrer (1950) and Roy (1956), involves‡ the combination of standardized aqueous solutions of metal nitrates or hydroxides with a suitable source of silica, which hydrolyses under certain pH and temperature conditions, giving a mixture with the consistency of a gel. This is then fired to convert nitrates and hydroxides to oxides and to remove undesired compounds used in the gelling process.

A number of variations of this generalized procedure have been published (e.g. Hamilton and MacKenzie 1960; Luth and Ingamells 1965; Hamilton and Henderson 1968; Biggar and O'Hara 1969). The two principal modifications are in preparation of the metal nitrates and hydroxides and in the source of silica. The metal nitrates and hydroxides are prepared either gravimetrically, by weighing suitable salts and converting them to the nitrate or hydroxide, or volumetrically, by adding them as prestandardized nitrate or hydroxide solutions. The silica may be added volumetrically or gravimetrically as an organic silicate, commonly tetraethylorthosilicate ((C_2H_5)SiO_4, TEOS), with ethyl alcohol added to ensure miscibility with the aqueous metal salt solutions; or it may be added as a prestabilized silica sol.

Equipment required. Very little specialized equipment is required for making gels. Because their preparation does not require heating above about 1000 °C, platinum-wound furnaces are unnecessary. However, to avoid contamination a platinum-basin is recommended for the final firing and platinum-crucibles for determination of silica yields of TEOS. During preparation the number of containers used should be kept at a minimum to avoid unnecessary loss caused by material adhering to the sides of the container, etc. This may be avoided by using 'Teflon' or polyethylene containers, from which the gel may be readily removed. Gels should never be made in less than 5-g batches, and preferably in 10-g lots. This is particulary important if TEOS is used as a silica source, since this volatile liquid is difficult to weigh accurately.

† The term *gel* is used to denote an amorphous solid which gives no X-ray diffraction pattern, and contains no birefringent material.

‡ Roy (1956) describes three types of gels. (a) Those for simple binary systems in which salts of two oxides are co-precipitated by each other at a predetermined pH. For those involving silica, care must be taken to remove adsorbed Na^+ and K^+ ions from the gel. (b) The simultaneous precipitation gel in which metal hydroxides and/or nitrates are mixed in solution, filtered, washed, and dried. These are suitable for mixtures involving more than two ions, particularly trivalent and small equivalent ions. Their limitations lie in their restriction to certain metals, and in difficulties with the sizes of crystals precipitated, owing to varying solubilities. (c) The organic silica-nitrate method is the one most generally applicable and will be referred to here simply as 'the gel method'.

Sources of components of gels. Gels containing a large number of different elements have been prepared. As in the previous section, only the sources and preparation for the principal rock-forming oxides are given. Sources of the less common elements are listed in Table 3.2.

(1) SiO_2. Two sources of SiO_2 are commonly used, TEOS or a soda- or ammonia-stabilized silica solution, sold under the trade name of 'Ludox' (E.T. du Pont de Nemours Company).

TEOS is a clear, colourless liquid, obtainable in a high degree of purity (Monsanto Chemicals, K and K Laboratories Inc.) and generally giving yields of 99·7 per cent or higher of SiO_2 on precipitation. Impurities are probably ethyl acohol. Hamilton and Henderson (1968) report no detectable sodium or potassium in the precipitated SiO_2. Before use, each bottle of TEOS must be standardized by weighing a known amount into a platinum crucible with lid, adding an approximately equivalent volume of absolute ethyl alcohol and gelling the solution by adding a few millilitres of NH_4OH (0·88 density), constantly stirring with a platinum spatula or 'Teflon' rod until a stiff porridge-like consistency is obtained. The gelling should be completed in a minute or two (in the presence of certain elements, particularly aluminium, this process is much more rapid). Material on the spatula or rod is washed into the crucible with ethyl alcohol, and wiped with a small piece of ashless filter paper, which is added to the crucible. The crucible is covered, allowed to stand overnight at room temperature to precipitate completely, and then dried on a water-bath or oven at 110 °C. The crucible and contents are finally heated to constant weight, and the SiO_2 yield obtained by weight differences. Initial heating must be gentle in order to fire the filter paper, which should not come into contact with the sides of the crucible, and to remove any excess NH_4OH or ethyl alcohol. Determinations should be made in triplicate, and the average yield used for the 'conversion' factor from $(C_2H_5)_4SiO_4$ to SiO_2. Periodic checks of the SiO_2 yield of each bottle of TEOS are advisable.

Because of its high volatility, TEOS is difficult to weigh accurately, but, fortunately, it loses SiO_2 stoichiometrically (Hamilton and Henderson 1968) and any loss can be allowed for prior to gelling. Although TEOS absorbs water and tends to hydrolyse if left unstoppered for long, Hamilton and Henderson report no percentage differences in yields of TEOS allowed to stand for 3 hours before being reweighed and gelled.

'Ludox' is a cloudy, ammonia- or soda-stabilized aqueous liquid containing particles of silica sol. In the ammonia-stabilized variety, the principal impurities are Na_2O, ZrO_2, and M_2O_3 (mainly Al_2O_3). 'Ludox' contains about 30 per cent SiO_2 by weight. Unfortunately, the amount of alkali impurity is very variable from batch to batch, and can be as high as 0·4 per cent by weight (Hamilton and Henderson 1968). Because of its variable Na_2O content and owing to possible concentration of SiO_2 and Na_2O by evaporation of water, 'Ludox' must be standardized before making each new batch of gels. It is not recommended for gels containing no Na_2O. For soda-bearing compositions a correction must be made for the Na_2O content.

Luth and Ingamells (1965) suggest the following procedure for standardizing

'Ludox'. A suitable aliquot of 'Ludox' is carefully measured into a platinum crucible, and slowly evaporated to dryness, avoiding loss by spattering or other causes. The crucible is ignited to constant weight, and the SiO_2 content is determined gravimetrically. Impurities in the residue may be determined by spectroscopy and flame photometry. The SiO_2 content of 'Ludox' cannot be determined by digesting the dried residue with H_2SO_4 and HF and weighing as SiF_4, since any alkali in the residue gives an erroneous value for SiO_2. It is also extremely important that the aliquot of 'Ludox' used for standardization be measured in exactly the same way as that used in the preparation of the gel itself, and preferably using the same burette or pipette.

(2) Al_2O_3. Two sources of Al_2O_3 are available, high-purity (99·99 per cent) aluminium 'dust', as described on p. 44, and $Al(NO_3)_3 \cdot 9H_2O$. Because of its deliquescence, the nitrate must be used as a carefully standardized solution. The standardization is done gravimetrically, and impurities are determined spectrographically, in a manner similar to that described above for 'Ludox'.

(3) K_2O. 'Spectroscopically' pure or 'Analar'-grade K_2CO_3 is used as a source of K_2O. The carbonate should be dried at 110 °C, and kept in a desiccator.

(4) Na_2O. 'Spectroscopically' pure or 'Analar'-grade Na_2CO_3 is a suitable source of Na_2O. This should also be dried and desiccated before use.

If the volumetric method of making gels is used, both K_2CO_3 and Na_2CO_3 are converted to their nitrates by carefully weighing the salt into a volumetric flask and adding dilute HNO_3. Owing to the rapid evolution of CO_2, the acid must be added slowly. The solution is then made up to the required volume with distilled and de-ionized water. As alternate sources of K_2O and Na_2O, Luth and Ingamells (1965) suggest CO_2-free, 1N aqueous KOH and 1N NaOH solutions available commercially as standardized volumetric concentrates.

(5) CaO. 'Spectroscopically' pure or 'Analar'-grade $CaCO_3$ is a suitable source of CaO. This should be dried and desiccated as described for glass preparation. If the volumetric method is used, the dried $CaCO_3$ is converted to the nitrate and diluted with distilled and de-ionized water by a method similar to that for the preparation of the alkali nitrate solutions. An alternative, but less recommendable, source of CaO is $Ca(NO_3)_2 \cdot 2H_2O$. If this is used, the nitrate solution must be standardized by EDTA titration or gravimetrically by precipitation of calcium oxalate and ignition to $CaCO_3$ or CaO.

(6) MgO. 'Pure' magnesium metal obtainable in 99·999 per cent purity (Johnson, Matthey Ltd.) is used as a source of MgO, although a slightly less pure form of the metal may also be used. In either case, it should not be purchased in any form that requires cutting or filing, because of possible contamination. For the volumetric method, a nitrate solution is made by a method analogous to that described under K_2O, Na_2O, and CaO. An alternative, but less recommended, source of MgO is $Mg(NO_3)_2$. Because of its deliquescence, the solution must be standardized by EDTA titration or gravimetrically by precipitation of magnesium ammonium phosphate and igniting

to $Mg_2P_2O_7$. Determination of impurities may be made by spectrographic analysis of the pyrophosphate.

(7) FeO. In the preparation of gels, FeO is usually added in the ferric state and reduced by passing hydrogen over the gels at about 600 °C for 2 hours (Ernst 1962), producing metallic iron. Such compositions are run using buffering techniques to control $f(O_2)$ as described in Chapter 8.

(8) Fe_2O_3. The choice of a suitable source of Fe_2O_3 presents a difficulty. Ferric nitrate solutions readily decompose in the absence of excess HNO_3 to form basic iron salts which are difficult to redissolve; mixtures of ethyl alcohol and nitric acid are potentially explosive.

For the gravimetric method, Hamilton and Henderson (1968) recommend metallic iron, 99·5 per cent pure. In order to prevent decomposition of the ferric nitrate, the solution is not evaporated to dryness (see p. 58), and care must be taken to avoid a violent reaction when TEOS and ethyl alcohol are added. Biggar and O'Hara (1969, using a volumetric method, advocate 99·999 per cent iron 'sponge' as a source of Fe_2O_3.

For the volumetric method of making gels, Heald, Reecher, and Herrington (1969) use ferric benzoate, approximately $Fe(C_7H_5O_2)_3$, as a source of Fe_2O_3. This salt dissolves in reagent-grade NN-dimethylformamide (DMF), to give a stable solution. These authors suggest that ferric benzoate be prepared from 50 g of ammonium benzoate dissolved in 800 ml distilled water. To this is added a solution of 50 g $Fe(NO_3)_3.9H_2O$ in 1 ml concentrated HNO_3 in 80 ml water, producing a light-brown precipitate which is suction-filtered, washed with distilled water, and dried at 50 °C. Ferric benzoate solution is made by dissolving the desired amount in DMF, giving a dark brown solution which is filtered through two Whatman No. 42 filter papers until it passes through easily. To standardize this solution, absorb a known volume of the solution on ashless filter paper and ignite in a platinum crucible, converting the ferric benzoate to Fe_2O_3. Initial heating is done gently to burn off the organic material, followed by ignition to constant weight in a muffle furnace at 900 °C to 950 °C. The concentration of Fe_2O_3 in the solution is then calculated.

(9) TiO_2. A number of sources of TiO_2 have been used by various investigators. Roy (1956) suggests triethanolamine titanium or tetrabutyl titanate. More recently, Ito and Frondel (1967, 1968) have used solutions of titanium tetrachloride in 6M HCl and titanyl dichloride in 4M HCl. Ammonium titanyl oxalate, $(NH_4)_2TiO(C_2O_4)_2.H_2O$, can also be used as a source of TiO_2 and has the advantage of being easily soluble in water (W. S. MacKenzie, personal communication).

Preparation of gels. During the past few years a number of different methods of gel preparation have been published which are all more or less modifications of the original techniques described by Roy (1956). Three methods of gel preparation, labelled A, B, and C are given here. Method A is essentially volumetric based on the method described by Luth and Ingamells (1965); method B is gravimetric, and is based on the technique of Hamilton and Henderson (1968); method C is a combined gravimetric and volumetric method, as used by Biggar and O'Hara (1969). Although the final products are basically the same and their compositions are probably equally accurate, the three methods are described here in detail so that the reader can compare the techniques in the light of the equipment and other resources available. The length of preparation may also be important; method B is slightly more rapid than the other procedures.

Method A. The required amounts of each of the standard nitrate solutions described above are dispensed from a burette into a screwcap polyethylene container. The burettes used should be the same as those used for standardization of stock solution, and every effort should be made to ensure that exactly the same techniques are employed in the gel preparation as were used in the standardization process. Solutions of calcium, magnesium, and aluminium nitrates are first mixed together and enough reagent-grade HNO_3 is added to prevent formation of hydroxides when the alkali nitrates and 'Ludox' are added. During the additions the solutions are mildly agitated by a magnetic stirrer placed in the polyethylene container. Alkali nitrate solutions and 'Ludox' are then added, and the bottle is tightly capped and allowed to stand at room temperature for a few days until a gel has formed; the gelling time required depends on the composition.

After gelling, the bottle cap is removed and any gel left on the cap and sides is washed into the bottle with distilled and de-ionized water. The upper part of the bottle is cut away with a razor blade and the bottle is covered with a watch-glass. The gel in the bottle is then dried at 50 °C to 60 °C, either in a standard drying oven or in a box lined with aluminium foil and heated with a 250-W infra-red bulb. The temperature in the oven or box must not exceed 70 °C or spattering will occur. Drying requires about a day and should be slow enough to give a uniform crust to the resulting powder.

The powder is carefully transferred to a platinum basin and gently heated over a Meker burner until nitrogen oxide fumes are no longer visible. The powder is then transferred to an agate mortar and ground until no coarse particles remain. After being replaced in the platinum basin, the powder is reheated to decompose any remaining nitrates. The temperature of this final heating depends on the nitrates involved, but should not be excessive, otherwise partial crystallization or even partial melting may occur, producing inhomogeneities in the gel and sites for nucleation of metastable phases in the experiments. The absence of such partially crystallized or melted gel may be periodically confirmed by removing a small portion of the material for examination by optical or X-ray-diffraction methods. After heating for $\frac{1}{2}$ to 1 hour, the gel is desiccated and ground to a fine powder in the agate mortar. As all gels are hygroscopic (and ones made by this method are extremely so), the powder must be stored in a tightly-sealed desiccator over anhydrous magnesium perchlorate.

Heald et al. (1969) describe a modification of this procedure for iron-bearing gels, using TEOS as a source of silica. This method involves pipetting the desired volume of each solution described previously into a fused silica dish, adding ten times the volume of water and two to three drops of concentrated NH_4OH. The dish is then covered to prevent evaporation and contamination of the TEOS, and dried for 3 to 5 days at 50 °C, or until gelling occurs. In mixtures with small amounts of TEOS, the dish is sealed to avoid evaporation of dimethylformamide and consequent crystallization of ferric benzoate before complete gelling. The desiccation and ignition procedure is the same as for non iron-bearing gels. For systems containing both soda and iron, 'Ludox' may be used as a silica source.

Method B. The required amounts of each pre-dried and desiccated constituent of the gel is weighed on a tared watch-glass. Each powder is carefully transferred to a 'Teflon' beaker. Sufficient dilute (1:1) HNO_3 is added to convert the carbonates and metals to the corresponding nitrates. This process is very rapid for magnesium metal and carbonates, but very slow for aluminium metal. The acid should be dispensed slowly from a squeeze bottle and a watch-glass should be placed on the beaker until the reaction has ceased. In order to dissolve the aluminium metal, the beaker is heated to about 70 °C on a hot-plate or water-bath. Even using aluminium 'dust' this process requires about 24 hours for most aluminosilicate compositions. When all the aluminium is dissolved the watch-glass is removed and any adhering liquid is carefully washed into the beaker with distilled and de-ionized water. The solution is evaporated almost to dryness, to remove excess HNO_3, and the residue is redissolved in a small volume of distilled and de-ionized water. If a large excess of HNO_3 has been added, this process should be repeated, because any excess HNO_3 may produce potentially explosive mixtures when the TEOS and ethyl alcohol are added. The final nitrate solution should be of small volume (about equivalent or less than the required TEOS) or difficulties in gelling may be encountered.

The required amount of TEOS is rapidly weighed in a crucible and carefully transferred into the beaker by washing with a sufficient volume of pure ethyl alcohol to produce completely miscibility between the TEOS and nitrate solution. This requires approximately equal parts of alcohol to TEOS and nitrate solution. While continuously stirring the mixture with a 'Teflon' rod, NH_4OH (0·88) is added until a stiff gel, containing no supernatant solution, is obtained. The time taken for gelling varies with the composition, but is normally less than a minute.

The beaker is covered with a watch-glass and allowed to stand overnight at room temperature, to ensure complete precipitation. The gel is then dried on a hot-plate at about 60 °C to 70 °C, followed by oven drying at 110 °C. During this process, requiring about 1 to $1\frac{1}{2}$ days, the rod remains in the beaker. When completely dry, the gel is easily removed from the beaker and rod, and is ground in an agate mortar. The resulting fine powder is transferred to a platinum basin, and is gently heated until all fumes of nitrogen oxides disappear. Finally, the gel is fired in a furnace at 800 °C (or some other suitable temperature) for a few hours, to convert all remaining nitrates to oxides. After cooling, the gel is checked for any sintering (if this has occurred, the gel must be reground). It is then bottled and desiccated, prior to use.

Weighed or titrated 'Ludox' may be used as a source of silica in this method. In this case, the 'Ludox' is washed into the nitrate solution with distilled and de-ionized water, and either allowed to gel slowly (requiring hours or even days) or gelled immediately by the addition of NH_4OH. For iron-bearing gels prepared by this method iron metal is used as a source of Fe_2O_3, and the nitrate solution is not completely evaporated to dryness before addition of TEOS to avoid decomposition of the ferric nitrate. Occasionally, it is necessary to use components of gels which are hygroscopic; they are weighed rapidly, but accurately, without regard to their calculated amounts, and the weights of the other components are adjusted to give a gel of the required composition.

Method C. This method has been used only for gels containing CaO, MgO, Al_2O_3, Na_2O, SiO_2, and iron oxide. According to Biggar and O'Hara (1969) it is unsuitable for gels containing either iron or magnesium and aluminium.

Nitrate solutions are prepared by weighing the required amounts of dried $CaCO_3$, magnesium metal, aluminium 'dust', iron 'sponge', and Na_2CO_3 into a volumetric flask along with a small volume of distilled and de-ionized water. The flask is placed in an ice-bath and 45 per cent HNO_3 is slowly added, using a capillary tube; this reduces the speed of the reaction of $CaCO_3$, Na_2CO_3, and magnesium with the HNO_3. To dissolve the iron and aluminium metals, the solution is heated to 70 °C to 90 °C for about 2 days, before diluting to the required volume with distilled and de-ionized water.

The required amount of each solution is weighed into a polyethylene container fitted with an airtight lid. To verify the weighings, the volume of solution dispensed by a burette is recorded. About 40 ml of pure ethyl alcohol is added to the solutions, followed by the correct weight of TEOS and are thoroughly mixed. The mixture is gelled by rapidly adding an excess of NH_4OH (0·88). Immediately after formation of the thick gel, the airtight lid of the container is replaced and the mixture is left at room temperature for at least 16 hours, to complete the gelling process.

With the rapid addition of excess NH_4OH, iron, aluminium, and magnesium hydroxides form first as gelatinous precipitates, followed by the hydroxides of calcium and sodium and, finally, silica gel. This produces homogeneity and prevents 'clots' of iron and aluminium hydroxides forming, which are difficult to disperse by agitation of the gel.

After gelling is completed, the gel is transferred to a 'Teflon' beaker by carefully washing with a water–ethyl alcohol mixture; the gel is dried for 2 to 4 days at 75 °C. The dried gel is then placed in a weighed platinum basin and slowly heated at 200 °C to remove most of the ammonium nitrate. This is followed by successive 2-hour heatings at 300 °C and 500 °C. Removal of volatiles and nitrates is accomplished by a final heating at 900 °C for 2 hours. At this point, the gel is cooled and weighed to give an indication of the yield. For a 10-g gel, Biggar and O'Hara (1969) regard a 9·97 g to 10·01 g yield as acceptable for iron-free compositions. For compositions containing iron, the yield depends on the oxidation state of the iron. Finally, the gel is ground for 10 minutes, and may be used in this form, or reheated for 48 hours at some temperature slightly below the temperature at the beginning of melting to give a partly crystalline product as a starting-material. In either form it should be desiccated until required.

Determination of gel compositions. Unfortunately, there is no easy physical method similar to that available for glass starting-materials for determining either the homogeneity or purity of gels. Provided the final heating has been below 1000 °C, preventing the formation of most mineral phases, gels of the common oxides prepared by these methods should be amorphous; they should show no birefringence under the petrographic microscope, and give no X-ray diffraction pattern.

Considerable controversy has arisen about methods for testing the homogeneity and compositions of gels (cf. Hamilton and Henderson 1968). Some authors (e.g. Shaw 1963; Luth and Ingamells 1965; Biggar and O'Hara 1968) report analyses of gels which are in very close agreement with their theoretical oxide contents. However, as Hamilton and Henderson (1968) point out, agreement between chemical analysis and theoretical composition of a gel does not necessarily imply that it is 'on' composition, unless it can be established that the gel is completely anhydrous. In many gels complete dehydration is very difficult, even after firing at 1000 °C. Hamilton and Henderson (1968) advocate that gels should be tested for homogeneity rather than determining their absolute composition. They argue that, if the yield of a gel (made up in the manner described as method B) is close to 100 per cent (in actual fact this should be slightly higher owing to absorbed water) and if the gel can be shown to be homogeneous, then the gel must be 'on' composition. This argument depends, of course, on the purity of the components of the gel, including the source of silica, the number of containers used in its preparation, etc. These authors describe a method for determining the homogeneity of gels using radioactive tracers, and conclude that gels made by their method (method B) are homogeneous, even before the grinding and mixing process.

Unlike glasses, in which an 'off' composition can be detected by irregularities in the liquidus curve, gels are often used for sub-solidus studies in which 'irregular' compositions are less easily detected.

The problem of ascertaining the composition and homogeneity of a gel is difficult. To minimize this problem the following precautions should be taken.

(a) The purest components available should be used and use of an excessive number of containers during preparation should be avoided, thus eliminating possible preferential loss of material.

(b) The final product should be examined both by optical and X-ray diffraction methods.

(c) The yield of the gel should be determined and compared with the calculated yield.

(d) All available methods should be used to determine both the composition and homogeneity of the gel.

(e) For many pure compounds both the temperature and sharpness of melting of gels can be compared with those of the same compositions prepared by other methods.

Dry and wet mixtures

Many experimental studies are done with mixtures of oxides or other compounds finely ground by dry or wet methods. Although this method is very acceptable for some compositions, for others it is completely useless, because the components of the mix are too unreactive, or because metastable phases are produced. As an example of this type of problem, Ehlers (1953) used mixes to study the system Al_2O_3–SiO_2–H_2O and, although the components of the mixes were chosen with care, he found no reaction between quartz and corundum after 60 days at 400 °C and $p(H_2O)$ of 10 000 lb in^{-2}. Although this is rather an extreme example, it illustrates one of the basic problems of preparing starting-materials as mixes.

Many investigators use mixtures of both pure chemical and natural minerals as starting-materials with satisfactory results. Use of natural materials is discussed in the last section of this chapter. Another technique is to make mixtures of the desired components, and fuse or sinter them to produce a homogeneous mixture which is then reground before use. Alternatively, mixtures can be made by suitable precipitation of the finely ground components. Whichever technique is used, the components of the mixes must be as reactive and as pure as possible, and the mixture must be homogeneous. One problem which may arise is that excessively reactive components may produce metastable phases; for some compositions it may be necessary to avoid certain very reactive components, e.g. α-cristobalite, which is known to persist metastably well below its equilibrium stability field.

Sources of components of mixes. The components of mixes are many and only a few of the commoner ones are listed here. Their methods of preparation are various and depend partly on the other components of the mix and partly on the specific problem under investigation.

(1) SiO_2. For experiments in which equilibrium is to be attained α-SiO_2 (quartz) cannot be used as a source of SiO_2. Some sources of SiO_2 are:

(a) pure quartz converted to cristobalite, as described in the preparation of glasses;
(b) reagent-grade silicic acid $SiO_2.nH_2O$ heated to 1200 °C for at least 24 hours;
(c) fused silica glass, crushed to fine grain size; and
(d) a commercial silica such as Cab–O–Sil (Godfrey L. Cabot, Inc.), heated to above 1000 °C to remove moisture, and slightly sintered to facilitate mixing.

Whichever source of silica is used, analysis should be made to ensure its purity.

(2) Al_2O_3. Because of its unreactivity, α-Al_2O_3 is unsuitable as a source of Al_2O_3. Various reagent-grade salts of aluminium, for example, $AlCl_3$, $Al_2(NO_3)_3$, etc., may be converted to γ-Al_2O_3 by first converting the chloride, nitrate, etc., to the oxide and heating at 1200 °C for 24 hours. As most salts of aluminium tend to be impure, the salt may be made from pure aluminium 'dust', as previously described.

(3) K_2O and Na_2O. Suitable sources of these compounds are difficult to obtain, because of their tendencies to form K_2O_4 and Na_2O_4 when exposed to air, and also because of their extreme hygroscopic natures. Probably the best sources are $K_2Si_2O_5$ and $Na_2Si_2O_5$, whose preparation has been described for glasses.

(4) CaO. Two sources of CaO are common. One is pure $CaCO_3$, dried at 500 °C for 5 hours, followed by heating for successive 2-hour periods at 1000 °C and 1250 °C to decompose the carbonate; alternatively, freshly opened reagent-grade $Ca(OH)_2$, heated for 2 hours at 240 °C may be used.

(5) MgO. Reagent-grade MgO should be heated for 1 hour at 1200 °C, or as described on p. 46.

(6) FeO. Reagent-grade $FeC_2O_4 \cdot 5H_2O$ or Fe_2O_3 can be used as a source of FeO. Both should be heated to 1000 °C for a few hours.

(7) Fe_2O_3. Pure iron 'sponge' or Fe_2O_3 can be used as a source of Fe_2O_3.

The components must always be kept desiccated until required. In order to produce a homogeneous mixture, components of mixes should all have approximately the same grain-size.

Preparation of mixtures. The preparation of mixtures is much simpler and less time-consuming than that of glasses and gels, because it involves only the accurate weighing of each component and the mixing of components, either by grinding under acetone or other suitable liquid, or by using a mechanical amalgamator† on the pre-ground components. Generally, the powder should not be too fine-grained, or disequilibrium phases will be produced arising from high surface-energies. A grain-size finer than 200 mesh but coarser than 325 mesh is probably sufficient to ensure homogeneity of the powder, but at the same time prevent metastability.

If desired, the mixtures may now be heated to a glass, at some temperature above their liquidus, or sintered, at some temperature just below their solidus. The glasses or sintered material are then ground and desiccated.

Determination of compositions of mixtures. Methods for determining compositions and homogeneites for mixes are similar to those for gels.

† 'Wig–l–bug' amalgamators, manufactured by the Crescent Dental Manufacturing Co., Chicago, is a recommended amalgamator.

They include fusing a mixture to a glass and comparing its refractive index to glasses of similar composition made by other methods and, particularly in the case of mineral compositions made as mixtures, determining the sharpness of their melting points.

Natural starting-materials

Natural minerals and rocks are used as starting-materials for experiments which do not involve a range of compositions in a given system. Recent developments in high-pressure techniques have permitted the investigation of the melting-relations and physical properties of minerals under mantle and even lower-mantle conditions using natural minerals. Valuable data on petrological processes have been gathered by determining the melting and crystallizing sequence of rocks in the laboratory.

The preparation of minerals and rocks as starting-materials requires little discussion. If minerals are used, the purest specimens should be chosen; after removing all extraneous impurities by suitable acid-leaching and washing, the mineral or rock should be analysed, and its composition precisely determined.

The chief difficulty in using rocks as starting-materials arises in ensuring that the small specimens chosen (usually only a few kilograms) are representative of the rock-type to be investigated, and that these specimens are homogeneous when further reduced to the much smaller sizes which are used in experiments (usually a few grams or milligrams). Ideally, one should start with several samples, each of which has been chemically analysed to determined their gross differences. Assuming, however, only one fairly homogeneous sample (e.g. a fine-grained non-porphyritic volcanic rock) of hand specimen size is available, it is sawn in half. One half of the sample is kept for optical examination, mineral separation, and modal analysis; the other half is ground to pass about 60 to 100 mesh in a stainless steel mortar, and any stray iron fragments are removed by a magnet and hand-picking. It is then thoroughly mixed by the method of 'coning and quartering', in which the powder is rolled on a piece of glazed paper or a clean rubber mat by tipping (in succession) alternate corners of the mat, until a homogeneous mixing is achieved. The heaped material is reduced in size by dividing it into four equal parts and removing diagonally opposite quarters. The remaining material is remixed, coned, and quartered, until the sample is reduced to a suitable size. This is then ground to about 200 mesh in an agate mortar, dried at 110 °C, and desiccated. During the last stage of this reducing process, the two quarters of the sample which are normally discarded should be analysed in the same way as the original sample, as a check on homogeneity.

Rocks which are inhomogeneous or contain minerals such as micas, which are difficult to grind, require special precautions. For this type of rock many samples should be available, and representative chip samples from each should be crushed and ground.

If required, the rock powder may be ignited and quenched to a glass before using.

Relative merits of different starting-materials

The relative merits of the four types of starting-materials must be considered in terms of the type of experiment being undertaken, the difficulties of preparation and homogeneity-testing, etc.

As mentioned above, natural materials are generally unsuitable for experiments requiring a range of compositions, but are extensively used for many high-pressure studies.

Glasses are particularly useful for liquidus studies, except those with high alkali compositions, in which loss by volatilization at the high temperatures required for their preparation may be serious. Glasses also tend to produce equilibrium results, since their free energies are much closer to those of the products than the free energies of other starting-materials, such as gels. This is fully discussed in Chapter 9. One disadvantage of glasses is that they may take a very long time to crystallize, owing to their high viscosity. This problem is particularly severe in the case of alkali-rich aluminosilicates.

Gels are useful for experiments in the sub-solidus region, because their high reactivity produces the desired assemblage in a reasonable length of time; these assemblages, however, may not be stable ones. Since it is unnecessary to heat gels to a high temperature during preparation, they are used for alkali-rich compositions in which the alkalis are not lost through volatilization. The fine-grained nature of gels may be advantageous in that it permits easy diffusion of ions in the structure during the experiments. Recently, Biggar and O'Hara (1969) compared the products of experiments in portions of the system CaO–MgO–Al_2O_3–SiO_2–Fe–O_2–Na_2O made with gels (by method C) and glasses of the same composition made by fusing the gels. They found that gels tended to retain metastable phases,† whereas glasses tended to fail to

---

† Shaw (1963) suggests that metastable liquid is more likely to form if the gel contains appreciable water.

nucleate stable phases. In most of their runs good agreement was found, using both types of starting-material, with respect to the number of phases present. Biggar and O'Hara could not decide which starting-material was more advantageous.

Experiments made with mixtures tend to produce metastable phases, or simply fail to react in a reasonable time, unless their constituents are carefully chosen.

With regard to their preparation and purity, glasses are tedious and expensive to make, but can be made up with very accurate compositions and good homogeneity. One distinct advantage is that their compositions and homogeneities can be verified by the simple optical method (see p. 47); the products of glasses are generally larger than those of gels, and they are therefore more amenable to optical identification. Glasses can also be kept for years and, if carefully desiccated, remain 'on' composition.

Gels are relatively easy and inexpensive to make, but determination of their compositions and homogeneity is difficult. Another difficulty is their tendency to nucleate metastable phases during their preparation; these are difficult to remove. The high reactivities of some of the components of gels, which may dissociate after several months or years, cause gels to go 'off' composition if kept for long periods. If the gel does not contain high alkalis or other volatile components it may readily be fused to a glass, and used as such. Indeed, for many compositions, this may be the most suitable form of starting-material, both from the viewpoint of ease of preparation and of homogeneity.

Some further considerations in the choice of starting-materials are given in Chapter 9.

References

BARRER, R. M. (1950). *Nature, Lond.* **166**, 562.
BIGGAR, G. M. and O'HARA, M. J. (1969). *Mineralog. Mag.* **37**, 198.
BOWEN, N. L. and SCHAIRER, J. F. (1932). *Am. J. Sci.* **24**, 177.
EHLERS, E. G. (1953). *J. Geol.* **61**, 231.
ERNST, W. G. (1962). *J. Geol.* **70**, 689.
ESKOLA, P. (1922). *Am. J. Sci.* **4**, 331.
FRONDEL, C. and ITO, J. (1968). *Am. Miner.* **53**, 943.
FYFE, W. S. (1960). *J. Geol.* **68**, 553.
GANGULI, D. and SAHA, F. (1967). *Trans. Indian Ceram. Soc.* **26**, 102.
GOLDSMITH, J. R. (1949). *Am. Miner.* **34**, 471.
GREIG, J. W. (1927). *Am. J. Sci.* **13**, 133.

HAMILTON, D. L. and HENDERSON, C. M. B. (1968). *Mineralog. Mag.* **36**, 832.
—— and MACKENZIE, W. S. (1960), *J. Petrol.* **1**, 56.
HEALD, E. F., REEHER, J. R. and HERRINGTON, D. R. (1969). *Am. Miner.* **54**, 317.
HIGNETT, T. P. and ROYSTER, R. H. (1931). *Ind. Engng. Chem.* **23**, 84.
ITO, J. (1968). *Am. Miner.* **53**, 890.
—— and FRONDEL, C. (1967). *Am. Miner.* **52**, 1105.
—— —— (1968). *Amer. Miner.* **53**, 1276.
KRACEK, F. C. (1930). *J. phys. Chem.* **34**, 2645.
—— (1939). *J. Am. chem. Soc.* **61**, 2863.
—— BOWEN, N. L., and MOREY, G. W. (1929). *J. phys. Chem.* **33**, 1875.
KUME, S. and KOIZUMI, M. (1965). *Am. Miner.* **50**, 587.
LUTH, W. C. and INGAMELLS, C. O. (1965). *Am. Miner.* **50**, 255.
MADORSKY, S. L. (1931). *Ind. Engng. Chem.* **23**, 78.
MOREY, G. W. (1951). *J. Soc. Glass. Technol.* **35**, 270.
—— and INGERSON, E. (1937). *Am. Miner.* **22**, 37.
PRESNALL, D. C. (1966). *Am. J. Sci.* **264**, 753.
ROEDDER, E. (1952). *Am. J. Sci.* A **250**, 435.
—— (1965). *Am. Miner.* **50**, 696.
ROY, R. (1956). *J. Am. Ceram. Soc.* **39**, 145.
SCHAIRER, J. F. (1951). in *Phase transformations in solids*. Wiley, New York. p. 278.
—— (1957). *J. Am. Ceram. Soc.* **40**, 215.
—— (1959). in *Physiochemical measurements at high temperatures* (ed. J. O. Bockris) Butterworths, London. p. 117.
—— and BOWEN, N. L. (1955). *Am. J. Sci.* **253**, 681.
—— —— (1956). *Am. J. Sci.* **254**, 129.
——, YODER, H. S. and KEENE, A. G. (1954). *Yb. Carnegie Instn Wash.* **53**, 62.
SHAW, H. R. (1963). *Am. Mineral.* **48**, 883.

4. Experiments at atmospheric pressure

Introduction

EXPERIMENTS at atmospheric pressure are relatively easy to perform, and need only simple easily maintained equipment. The majority of laboratory studies pertinent to silicate† petrology and ceramics have been done with this type of equipment, using the quenching method originally developed by Shepherd, Rankin, and Wright (1909). In this technique, the equilibrium state present at the temperature of the experiment is 'frozen in', and crystalline and liquid phases (as glass) can be observed optically or by X-ray diffraction methods. Because of the tendency of silicate glasses to devitrify rapidly, the quenching method is is the only reliable way of determining phase equilibrium in silicate systems. In addition to liquidus relations, many quenchable polymorphic inversions can be investigated by this method.

Although experiments at atmospheric pressure are only strictly applicable to the genesis of volcanic rocks and those formed at very shallow depths, almost all systems are initially investigated under atmospheric conditions. Such systems are often loosely referred to as 'dry' systems, to distinguish them from hydrothermal systems (see p. 7) and to indicate that water or some other volatile is not a component. However, experiments done under controlled atmospheres at a total pressure of approximately 1 bar using gas mixtures cannot be referred to as 'dry' experiments. These are described at the end of this chapter.

Because the furnaces and temperature-measuring devices used in 1-atmosphere experiments are the same as those used in experiments under pressure (described in later chapters), a fairly detailed account of the types of thermocouples and their calibration is given here. Owing to rapid advances in the field of electronics, only basic details of temperature controllers are included.

Temperature measurement

Temperature scales

In scientific work all temperatures are recorded in degrees celsius or degress kelvin. In contrast to parameters such as length and weight,

† The term 'silicate' is used here since many reactions involving sulphides are not quenchable.

which may be easily defined, temperatures must be defined on the basis of a set of experimentally reproducible phenomena such as melting-points or boiling-points. In 1914 workers at the Geophysical Laboratory set up a temperature scale (the Geophysical Laboratory Temperature Scale) based on nitrogen thermometry and with eleven principal fixed points. This scale, still employed at the Geophysical Laboratory, can be used from 0 °C to 1755 °C. It was not until 1927 that international agreement was reached on a temperature scale (International Temperature Scale of 1927). This scale was revised in 1948 (International Temperature Scale of 1948). Fig. 4.1 shows the comparison between these three scales.

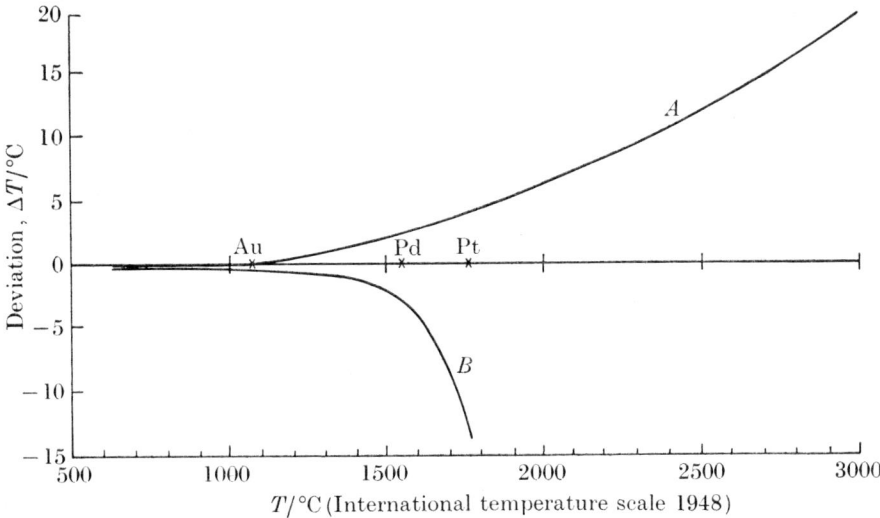

FIG. 4.1. Comparisons of various International Temperature Scales with Geophysical Laboratory Temperature Scale of 1914. Geophysical Laboratory Scale is shown as straight lines, A, $\Delta T = $ Int. 1927–Int. 1948; B, $\Delta T = $ Geophysical 1914–Int. 1948. (After Levin et al. 1964.)

From 0 °C to 1550 °C differences between the scales are negligible, and within the limits of error of most phase studies. However, above 1550 °C the differences are significant, and in this region the temperature scale being used should be noted in reporting data. Sosman (1952) has given an account of the development of the various temperature scales.

Methods of measuring temperature

For most experimental studies on silicates, sulphides, and other geological material, temperature is measured by a thermocouple or an optical pyrometer. Of these two methods, the thermocouple is by far the commoner; it can be used for temperatures from 0 °C to about 1600 °C. The optical pyrometer is generally only used for temperatures exceeding 1600 °C, outside the melting-range of most petrologically important materials. In addition, the accuracy of the thermocouple method is much greater than that of the optical pyrometer.

Thermocouples. A thermocouple consists of two dissimilar electrically-conducting metals joined at one end. This end is placed at the point where the temperature is to be measured (hot junction). The temperature differences between the joined and free ends of the thermocouple generates an e.m.f proportional to the temperature. This e.m.f. is measured by a potentiometer and converted to the corresponding temperature using standard reference tables of e.m.f. plotted against temperature for different thermocouple materials.†

The choice of thermocouple depends on a number of factors, including the temperatures expected, the accuracy required, the physical conditions of the experiment (e.g. the type of atmosphere employed), the composition of the thermocouple wires, and their possible contamination from the material being heated. A basic requirement of all thermocouples is that the wires must be chemically homogeneous and of uniform diameter. If these specifications are not met the e.m.f. will change along the length of the wire giving erroneous temperature readings. All new thermocouple wires must be calibrated, as described in a later section of this chapter.

A list of commonly used thermocouple wires and their ranges of temperature is given in Table 4.1. Of these, platinum–rhodium and chromel–alumel‡ have the most useful temperature ranges for petrological work. Below 100 °C Cr–Al thermocouples are generally used providing the atmosphere is such that the wires are not oxidized or otherwise attacked. Cr–Al wires tend to become brittle, particularly at high temperatures, and the thermocouple should be replaced after each run under these conditions. Cr–Al thermocouples show an average change of 0·04 mV per °C temperature change, and therefore even with a small portable potentiometer, temperatures can be determined to within 0·2 °C. Above 1000 °C, Pt–PtRh thermocouples are usually used.

† A recommended set of such tables is available from the Superintendent of Documents, U.S. Government Printing Office, Washington, D.C. as *Reference Tables for Thermocouples*, *National Bureau of Standards Circular* 561.

‡ Types of thermocouples are generally abbreviated. For example, a thermocouple in which one wire is pure platinum, the other 90% platinum 10% rhodium is abbreviated as Pt–Pt$_{90}$Rh$_{10}$. Similarly a chromel–alumel thermocouple is referred to as Cr–Al.

TABLE 4.1

Some common thermocouple wires and their temperature ranges†

Wires	Temperature range (°C)	Maximum temperature (°C)
Chromel–Alumel (90%Ni–10%Cr) (95%Ni, 5%Al, SiMn)	−200 to 1200	1350
Pt–Pt$_{90}$Rh$_{10}$	0 to 1450	1700
Pt–Pt$_{87}$Rh$_{13}$		
Pt–Pt$_{80}$Rh$_{20}$		
Copper–Constantan (60–45%Cu, 40–55%Ni)	−200 to 350	600
Iron–Constantan	−200 to 750	1000

† Data for this table have been taken from Roeser (1941).

In addition to their higher temperature range these thermocouples have the advantage that they are not readily attacked by the sample, except in the case of iron-bearing samples where, at high temperatures, iron alloys with the platinum. Similarly, it is important that the platinum be free from iridium, which produces volatilization with subsequent drift in readings at high temperatures. Pt–PtRh thermocouples show an average change of 0·012 mV per °C temperature change, and can thus be read with an accuracy of about 0·3 °C with a small potentiometer and more accurately with a larger instrument.

The diameter of the wires used in thermocouples depends largely on the experimental apparatus. If the investigator is making his own thermocouples, the thickness of wire is largely a matter of personal choice, but it must be such that the two wires are the same and can readily be joined by welding or some other method. Schairer (1959) recommends 0·4-mm diameter wire for Pt–PtRh thermocouples. Thicker wire should generally be used for high temperatures, to protect against breakage resulting from brittleness. The alternative is to use sheathed thermocouples in which the wires are completely enclosed in some suitable material, to protect them from mechanical damage. This type of thermocouple is very useful for experiments in which corrosive liquids or gases are present or for experiments in which the temperature is measured under pressure. One disadvantage of sheathed thermocouples is that their response to temperature changes is slower than thermocouples in which the welded tip is exposed.

A schematic diagram of a simple thermocouple circuit is shown in Fig. 4.2. The thermocouple wires are usually not attached directly to the potentiometer or other measuring device, but are connected to a terminal head (or thermocouple connection) located near the furnace This connects the thermocouple wire to some wire of suitable conductance, which in turn is attached to the reference junction† of the measuring instrument. These wires are known as *extension wires* or *lead wires* and are unique for each type of thermocouple. Temperatures are recorded either on a potentiometer (as an e.m.f. which is converted to

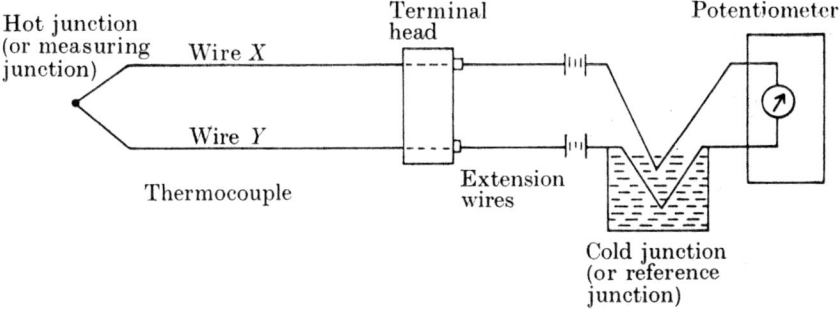

FIG. 4.2 Schematic diagram of simple thermocouple circuit.

degrees) or on a temperature-recording device consisting of a potentiometer and automatic recorder (such as a strip chart) which is calibrated directly in degrees. In the latter type of instrument many temperatures can be recorded continuously.

Construction and calibration of thermocouples. The construction and calibration of thermocouples is extremely important in all experimental work. Thermocouples are used both for measurement of the sample temperature and to operate temperature controllers. For both purposes the thermocouple wires must be insulated from any metal or other material which might alter their e.m.f. and, for the same reasons, they must not be allowed to come in contact with one another, except at the hot junction. Similarly, the thermocouple tip, if exposed, must not touch the sample, for it would thereby produce incorrect e.m.f. values.

Thermocouple wires, sufficiently long to reach from the hot junction to the terminal head, are threaded through a piece of double-bore ceramic tubing of sufficient length to prevent the thermocouple wires

† The reference junction (or cold junction), as opposed to the measuring junction (or hot junction), is the end of the thermocouple subjected to a known temperature. Many potentiometers have an automatic reference junction.

touching any part of the furnace. Mullite or alumina ceramic tubes are suitable for most thermocouples, and can be used up to 1900 °C. One end of each wire is welded together; for standard Cr–Al and Pt–PtRh couples this is done by twisting a short length of the wires around one another, and welding the two wires in a d.c. arc with a sharp pure-graphite pencil electrode (see Chapter 5). About 80 V are required for welding Cr–Al thermocouples and about 70 V for Pt–PtRh thermocouples. Any length of wire twisted together is unwound and drawn back through the ceramic tubing, leaving only the tip of the thermocouple exposed. Before proceeding, the welded tip should be examined under a microscope to ascertain that a good connection has been made. The thermocouple is then attached to the terminal head, observing the correct polarity of the wires.

Thermocouples are calibrated by determining the melting-points of substances which melt sharply at a known temperature. Such substances must be of the highest purity and must have melting-points covering the temperature range in which the thermocouple will be used. Common materials for thermocouple calibration and their melting-points are are given in Table 4.2. Preparation of some of these materials is given in

TABLE 4.2

Commonly used materials for thermocouple calibration

Material	Melting temperature (°C)	Material	Melting temperature (°C)
Zn	419·5	Li_2SiO_3	1201·0
NaCl	800·4†	Diopside	1391·5†
Ag	960·8	$CaSiO_3$	1544·0†
Au	1062·6†	Pd	1549·5†

† Used as calibrants at the Geophysical Laboratory. Temperatures quoted on Geophysical Laboratory Temperature Scale (Schairer 1959).

Schairer (1959). Zinc dust should be free from any oxidation or surficial impurities, and preferably of 'spectroscopically pure' grade. NaCl is prepared by a double precipitation of pure NaCl from pure concentrated HCl, followed by washing with HCl, drying, grinding, and a final drying for 24 hours at 100 °C. Schairer (1959) reports the melting-range of NaCl, prepared in this way, as negligible. Pure gold and palladium require no preparation other than cutting into thin slivers with shears or a lathe. Diopside ($CaMgSi_2O_6$) is synthesized from $CaCO_3$, ignited MgO and pure SiO_2, as described in Chapter 3. The crystallized glass of diopside composition has a melting-range of 3 °C to 4 °C, because of traces of alkalies in the MgO. An accurate temperature can be attained, however, by determining the temperature at which the last crystals melt. $CaSiO_3$ is prepared from pure $CaCO_3$ and SiO_2, and has a melting-range of less than 1 °C.

Two procedures may be used for calibrating thermocouples. The first method entails determination of the melting-point by a heating and cooling curve; the second method entails precise determination of the melting-point by the quenching technique. Using the second technique, thermocouples can be calibrated to within ± 0.4 °C, whereas the heating- and cooling-curve method produces an accuracy of only approximately ± 0.8 °C, depending on the substance used for calibration.

In the first method, the powdered calibrant is placed in a suitable container, usually a gold tube of length $\frac{1}{3}-\frac{1}{2}$ in, welded at one end, or an envelope made from platinum foil (see p. 85). The hot junction of the thermocouple is embedded in the powder, providing intimate contact between the two. For substances which are not easily powdered, such as gold and palladium, a short length (about 1 mm) of the sample to be melted is welded to the hot junction of the two thermocouple wires. The thermocouple is then placed in the predetermined hot spot of a newly wound furnace. The other ends of the thermocouple are connected by means of extension wires to a reference junction (consisting of a wide-mouthed Dewar bottle containing shaved ice saturated with water) and then to a potentiometer. The furnace is heated slowly and e.m.f. readings are taken at suitable time-intervals. As the temperature of the melting-point is reached, the e.m.f. remains constant for a short time (usually less than a minute). When melting is complete, the temperature again rises. The furnace is now cooled slowly, and a similar constancy in e.m.f. is observed as the sample freezes. A time-e.m.f. plot is then made for both heating and cooling processes, as shown in Fig. 4.3. The e.m.f. corresponding to the flattened portion of the curve is added to the e.m.f. corresponding to the temperature of the room for the particular thermocouple being used. This temperature is obtained from a thermometer capable of reading to ± 0.1 °C, placed on the potentiometer. The combined e.m.f. values are then converted to temperature, using the appropriate conversion table. This value should correspond closely to the accepted value. Any deviation should be noted, and the appropriate correction made when using the thermocouple.

74   *Experiments at atmospheric pressure*

There are two possible sources of error in this method. If heating or cooling is rapid, the transition temperature may be passed too quickly for any constancy in e.m.f. values to be observed. A second source of error is contamination of the thermocouple by the sample, but this is generally not significant because the two are in contact for only a short time.

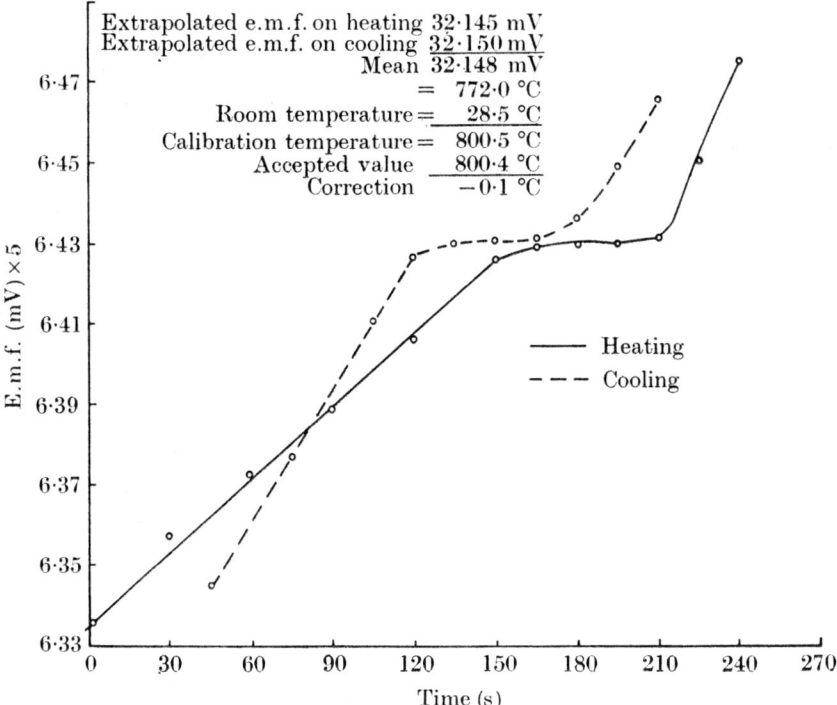

FIG. 4.3. Time–e.m.f. plot for calibration of thermocouple wire by heating and cooling method using NaCl as calibrant.

The second calibration-technique involves the quenching method, described in detail later in this chapter. To determine the exact melting-points of powdered material (zinc, NaCl, diopside, and $CaSiO_3$), a small portion of the crushed powder is placed in a platinum envelope and suspended on the quenching rig (described on p. 80 and shown in Fig. 4.4(b)). The hot junction of the thermocouple is placed about 1 mm from the platinum envelope, and both are placed in the hot spot of a newly-wound quenching furnace (Fig. 4.4(a)). This is precisely the same procedure as is used in quenching experiments, and thus eliminates any significant temperature differences arising from inhomogeneities in the thermocouple wire or differences between the hot junction and sample.

If gold or palladium is used as calibrant, a sliver of the metal is placed in a small-bore porcelain tube, plugged at both ends with asbestos, and the tube is placed in a platinum envelope of the same size as is used for the powdered sample, thus preventing the thermocouple from coming into contact with the sample. The arrangement of extension wires, reference junction, and potentiometer is otherwise the same as in the first method.

The furnace temperature is slowly raised to a value slightly lower than the known melting temperature, and the sample is quenched in air or mercury, as described on p. 85. If unmelted, samples remain as powders and easily fall from the platinum envelope, leaving a clean surface. For diopside and $CaSiO_3$, microscopic examination in oil of suitable refractive index shows the presence or absence of crystals. Unmelted metallic samples remain as slivers in the porcelain tube, melted powders appear as glazed glass filling the interstices of the platinum envelope, and melted metals occur as globules in the porcelain tubes.

The temperature of the furnace is raised or lowered by an amount corresponding to 10 μV (according to the results in the first heating), and the experiment is repeated until the melting-point is 'bracketed' to ± 5 μV (corresponding, in the case of Pt–Rh thermocouples, to about 0·4 °C). The e.m.f. of the reference junction is added to the determined melting temperature, and deviations from accepted melting temperatures are obtained from reference tables, as in the previous method.

Thermocouples may also be calibrated by comparison with standard thermocouples, and by other methods; the reader is referred to Roeser and Wensel (1935) for these procedures.

The frequency of calibration of thermocouples and the number of calibrants depends largely on the type of thermocouple and on the temperature range for which it is used. In the upper part of their temperature range, Cr–Al thermocouples become brittle owing to oxidation and should be changed after every experiment. In contrast, Pt–Rh thermocouples may be used many times, provided they are not exposed to contamination, particularly from alkalis and iron at high temperatures. If this happens, the thermocouple should be recalibrated immediately. Otherwise calibration three or four times a year is sufficient for thermocouples used in the range of 700 °C to 1300 °C. Above 1300 °C, calibration may change by as much as 1 °C to 2 °C per week (Schairer 1959), and thermocouples should be calibrated frequently. For accurate measurements above 1450 °C, calibration should be done before and after each experiment, using diopside and $CaSiO_3$. The number of calibration points depends on the temperature range and the accuracy required. For the most precise measurements, the thermocouple should be calibrated at as many fixed melting-points (Table 4.2) as are available.

The problem of thermocouple calibration and the comparability of results between different laboratories has been discussed by Biggar and O'Hara (1969). Using a technique described later in this chapter, these authors, using six approximately equally spaced calibration points, showed that the temperature corrections in the range 800–1600 °C were not linear. On the basis of these results they suggest that temperatures of previous work, in which fewer calibration points were used, may be in error by as much as 10 °C near 1200 °C. Biggar and O'Hara also used reaction or eutectic temperatures in calibration rather than melting-points, since the former do not require precise control of the ratios of the major components involved. They state that the main factors influencing the comparability of results are the temperature scale used, the number of calibration points, the purity of the calibrants, the proximity of experimental temperatures to calibration points, the experimental method used, and the type of phase change involved in the calibration technique.

The optical pyrometer. The optical pyrometer measures temperature by determining the brightness of colour (for a given wavelength) of the surface of a body. Such a determination depends on relating brightness to temperature when referred to a standard source, usually a black body.† The theory and construction of optical pyrometers are given by Forsythe (1941) and Margrave (1959). Complexities in construction and difficulties in calibration make the optical pyrometer useful primarily for temperatures in excess of 1600 °C, where most thermocouples are unsuitable. Although the optical pyrometer may be used at temperature up to 3700 °C, its accuracy is considerably less than the thermocouple as a method of temperature measurement. Margrave (1959) states that the reproducibility of optical pyrometers when used by an experienced operator is only $\pm$ 2–3 °C at 1000 °C, and is probably less at higher temperature.

Furnaces

Many different types of furnace may be used for experiments at atmospheric pressure. In this section some general considerations, such as

† A black body is a radiator which, for a given wavelength, emits the maximum energy per unit of time at any specified temperature. The energy given off is a function only of temperature for each part of the spectrum.

Experiments at atmospheric pressure 77

choices of metals for windings, furnace tubes, etc., are given. As examples, descriptions of the standard quenching furnace and a temperature gradient furnace are presented. Both of these may be modified for specific purposes.

Windings

A resistance furnace consists of a wire of a suitable conducting material wound on a ceramic tube. As discussed in Chapter 3, (p. 41) furnace elements are also made as spiralled tubes. The choice of conductor depends on:

(a) the temperature desired;
(b) its temperature coefficient of resistance;
(c) its chemical characteristics;
(d) its ductility, brittleness, coefficient of linear expansion, etc.

For temperatures below approximately 1250 °C there are a number of non-noble metal wires suitable for furnace windings. These wires are base metal alloys of iron, chromium, aluminium, nickel, cobalt, silicon, carbon, and other elements; they are sold under a variety of trade names. Each wire has slightly different properties, and the choice of wires depends on the maximum temperature required and the specific atmosphere in the furnace. For quenching experiments under atmospheric conditions, the following wires are extensively used: 'Nichrome' V,† suitable to 1150 °C; 'Alferon',† suitable to 1225 °C; 9nd 'Kanthal' A,‡ suitable to 1330 °C For temperatures from about 1250 °C to 1650 °C, pure platinum or a Pt–Rh wire is usually used. Below about 1600 °C, silicon carbide can also be used. Because of the tendency of platinum to alloy with iron at high temperatures, platinum windings should not come into contact with iron in the samples. Platinum also alloys with aluminium, and in certain proportions these two elements may melt at a very low temperature (about 600 °C). For temperatures in excess of 1650 °C, molybdenum or tungsten windings may be used. Molybdenum can be used up to about 2400 °C, but readily oxidizes at high temperatures. Tungsten windings can be used to over 3000 °C, but rapidly become brittle.

After selecting a suitable wire for the temperatures desired, the cross-sectional diameter of the wire, the diameter of the coil, and the length of wire are calculated. Although the wattage input determines the temperature, if the diameter of the wire is increased the operating temperature is

---

† Trade marks of Driver–Harris Company.
‡ Trade mark of Kanthal Corporation.

decreased. The maximum wattage depends on the amperage and resistance, according to Ohm's law:

$$\text{watts} = (\text{amperes})^2 \times \text{ohms},$$

or watts = volts × amperes, where volts = amperes × ohms.

From a knowledge of the available voltage (E) and the desired power (P) in watts, the hot resistance (R_h) is calculated using the formula

$$R_h = E^2/P \qquad (4.1)$$

and the cold resistance (R_c) is calculated using the formula

$$R_c = R_h/C_t, \qquad (4.2)$$

where C_t is the temperature coefficient of resistance normally supplied by the wire manufacturer, and approximately equivalent to $1 + \alpha T$, where α is the coefficient of thermal expansion of the wire and T is the desired temperature.

For example, consider the length of wire required in building a furnace for a 115-V supply to deliver 800 W using 'Nichrome'-V wire. The desired temperature of the furnace is about 1000 °C. From eqn (4.1), R_h is 16·53 Ω whence, from eqn (5.2), R_c is 15·44 Ω (C_t being determined from the manufacturer as 1·07 for 1000 °C). For 15·44 Ω cold resistance, the correct wire gauges are 18 to 22 B and S. If 20-gauge wire is selected, it resistance is 0·659 Ω ft^{-1}, as determined from a standard wire-resistance table (e.g. *Handbook on physics and chemistry*). The total length of wire required is then

$$\frac{\text{cold resisance } (R_c)}{\text{resistance of wire}} = \frac{15\cdot44 \text{ Ω}}{0\cdot659 \text{ Ω ft}^{-1}} = 23\cdot4 \text{ ft}.$$

The total number of coils required is determined by the mean diameter of the coil (tube diameter × wire diameter) and the total length of the wire, using the formula

$$\text{number of coils} = \frac{\text{total length of wire}}{\pi(\text{mean diameter of coil})}.$$

Thus, if the mean diameter of the coil is 1 in, the number of coils required

$$\text{is} \quad \frac{23\cdot4 \times 12}{1 \times 3\cdot14} = 89.$$

The number of coils may be evenly spaced over the length of the furnace tube, preferably with approximately one wire thickness left between each coil, or the number of coils per inch may be varied. When calculating the length of wire required for furnace windings, allowance should be allowed for attaching the windings to a binding post on the outer furnace shell, from which leads connect it to the power supply or temperature controller.

Most wires used for furnace windings are heat-treated by the manufacturer and can be used directly. This treatment effects their ductility, brittleness, etc.; it thus prolongs their life and makes winding on the ceramic furnace tube easier. Platinum wire should be preheated to 600–700 °C for a few hours. The wire is most easily wound on the tube using a lathe. By anchoring the wire at one end and setting the lathe at the correct speed, the wire can be evenly wound round the tube at the desired number of turns per inch. The windings are then covered with a suitable cement (e.g. 'Alundum' RA-98†) and allowed to dry before baking to harden.

Ceramic tubes

The choice of ceramic tubes for furnaces depends on the temperature, types of atmosphere, and rate of heating and cooling to be used. The commonest ceramics for furnaces used in experimental petrology are made of mullite or alumina. Various grades of both materials are available. Mullite can be used up to about 1760 °C, depending on its composition; it has a low thermal conductivity ($2 \cdot 5184$ $Wm^{-1}K^{-1}$), is impervious to air, hydrogen, and carbon monoxide, and has a low thermal expansion ($5 \cdot 0 \times 10^{-6}$ °C^{-1}). Pure mullite has a high chemical resistance at elevated temperatures and is resistant to thermal shock and 'sagging'. Alumina has a slightly higher thermal conductivity ($0 \cdot 050$ cal/s/cm^2/cm/°C) than mullite. It is resistant to reducing atmospheres and chemical attack at high temperatures. In addition to its suitability for higher temperatures, alumina is also considerably stronger than mullite.

Quenching furnace

The quenching furnace described here is a modification of the design used at the Geophysical Laboratory and described by Schairer (1959). Similar furnaces have been described by Faust (1936). This furnace is similar to the vertical making-furnace used in the preparation of glasses, which was described in Chapter 3.

Details of the furnace and quenching rig are shown in Fig. 4.4. The furnace consists of three coaxial tubes; the outer and inner tubes are

† Trade mark of Norton International Incorporated.

composed of mullite, the centre tube is of high-grade alumina. Dimensions of these tubes are shown in Fig. 4.4(a). Ninety-five grams of 0·8-mm pure platinum wire are evenly wound at 11 turns per inch on the centre tube, which is placed in a vertical position in a steel jacket, approximately 7 in in diameter and 10 in long. The top and bottom of the jacket are enclosed in cemented asbestos or 'Transite' board, with a hole in the top for centring the furnace tube and a hole in the bottom to permit quenching. The furnace tube is surrounded by the outer tube of slightly larger diameter and the space between the tubes filled with 60-mesh granular alundum to prevent local melting of the windings at high temperatures, which would cause failure of the furnace. The outer tube therefore prolongs the life of the furnace and provides a more even heat distribution. The inner tube is about $1\frac{1}{2}$ in longer than the other tubes and projects above and below the furnace. It also prolongs the life of the furnace by protecting the furnace tube from accidental leakage of material from the sample.

The space between the outer tube and steel casing is tightly packed with lightly-calcined MgO, which acts as an insulator. This must be frequently repacked during the early life of the furnace, because of its tendency to shrink on heating. The furnace is mounted on steel legs, leaving about 6–8 in at the bottom to allow room for the cup of mercury used for quenching. The bottom of the centre tube is plugged with silica wool.

Schairer (1959) states that quenching furnaces of this type have a life of several years if operated below 1200 °C, 4 to 6 months at 1400 °C, 3 to 4 weeks at 1500 °C, and only a few days at 1600 °C. Although not extensively used, similar quenching furnaces could be made with silicon carbide heating elements and insulated bricks. According to manufacturers' specifications, these are approximately equivalent to platinum-wound furnaces, and are slightly less expensive.

The quenching rig, containing the thermocouple and sample-suspension tube, is shown in Fig. 4.4(b). This is made of soapstone, machined to fit tightly on top of the centre tube. Two holes are drilled in this rig to hold the thermocouple and sample-suspension tube in such a manner that the thermocouple tip and sample are equidistant from the furnace wall but as close together as possible. The sample-suspension tube is made of double-bore ceramic, with bore size large enough to accommodate 0·7-mm platinum wire. The bottoms of these wires are bent to form a hook, upon which the sample is suspended by a 0·2-mm platinum wire, in such a way that it lies as close as possible to the thermocouple tip. The top ends of the suspension wire are connected to a

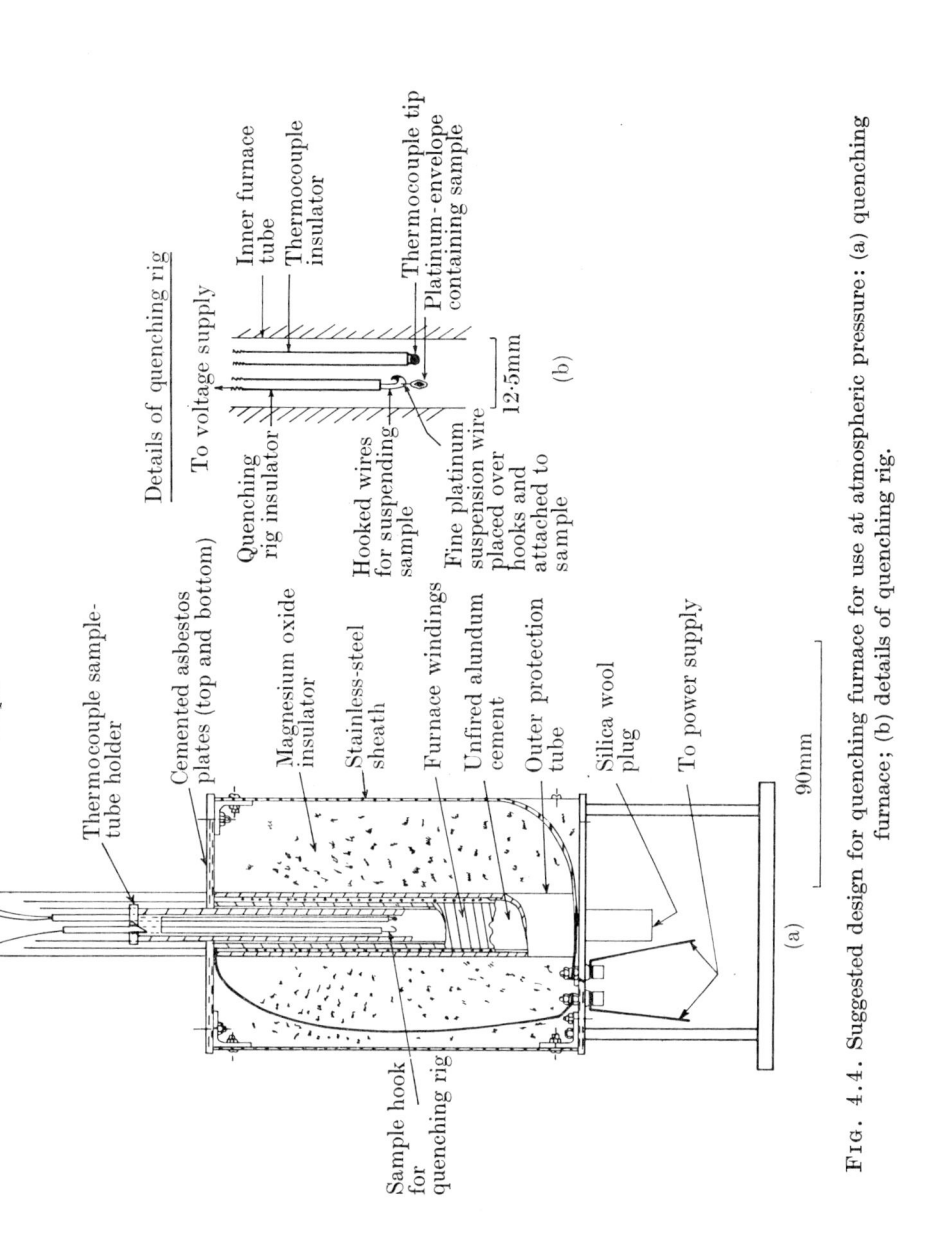

FIG. 4.4. Suggested design for quenching furnace for use at atmospheric pressure: (a) quenching furnace; (b) details of quenching rig.

transformer to quench the charge at the end of an experiment, as described on p. 85. Schairer (1959) describes a modification of this quenching rig in which the sample-suspension wires and thermocouple wires are placed in separate tubes, requiring a four-hole rig.

In any furnace of this type it is necessary to determine the centre of the 'hot spot', or zone of maximum constant temperature, for a given regulated temperature. For the quenching furnace described above, the 'hot-spot' is about 3 cm long. The temperature along the 'hot spot' should not vary by more than 0·5 °C. This zone can be located by selecting a convenient temperature or sets of temperatures and, by holding the thermocouple tip at various heights in the furnace until thermal equilibrium is achieved, determining the centre of the 'hot spot'. Lateral variations in temperature can also be recorded by modifying the quenching rig to contain thermocouples at the circumference as well as the centre of the tube. Details of these are given by Biggar and O'Hara (1969). During the life of the furnace, the 'hot spot' may shift (particularly if the apparatus is being used at temperatures above 1450 °C); this is caused by volatilization. The shift results in changes in the resistance of the platinum windings. The 'hot spot' should be periodically checked.

Thermal-gradient furnaces

Thermal-gradient furnaces are used principally for crystal-growth studies. Although such studies are normally done in a hydrothermal environment, the furnace design is similar for both dry experiments and experiments under pressure. For thermal-gradient furnaces used without an adjacent pressure-vessel, the temperature gradient depends on the arrangement of furnace windings and on the total length of the furnace. If pressure-vessels are used, the furnace windings may be in two or more independently controlled zones, and the furnace and pressure-vessel may be moved relative to one another to establish a temperature gradient along the vessel.

The windings for a thermal-gradient furnace are such that the square of the current density along the wire and the thermal insulation provide an almost linear temperature gradient. Thermal-gradient furnaces are normally operated in a horizontal position, because of the greater possibility of thermal convection in vertical furnaces. If a well-defined temperature is not desired, but only a thermal gradient along the furnace length, a single winding is sufficient. At the 'hot' end of the furnace the number of turns per unit length will be greatest, and will decrease toward the 'cold' end of furnace tube. For experiments in which a well-defined temperature-profile is necessary, a series of independently controlled windings can be made, with a different number of turns per unit length in each furnace-zone. In both types, thermocouples are placed along the length of the furnace to determine temperature-profiles. A typical temperature gradient for a two-zone 'Nichrome'-wound furnace might be 400 °C over a length of 10 in. In thermal-gradient furnaces it is particularly important that thermal insulation be uniform particularly at the ends to prevent localized heat loss.

Temperature controllers

In quenching furnaces temperature may be controlled by two methods.

The first method involves direct temperature regulation through the furnace windings. This can be done because of the direct relationship between the resistance and temperature of the wire, and the method is particularly suitable for platinum-wound furnaces, in which resistance changes rapidly with temperature. The furnace windings are attached to a variac or other suitable transformer. The transformer's voltage input is stabilized, preventing fluctuations in line-voltage from affecting temperatures. Commercially available voltage stabilizers will maintain temperatures at ± 1 °C for extended periods. For very accurate control, the method in use at the Geophysical Laboratory (Schairer 1959) may be used. This is based on the Wheatstone-bridge principle (Roberts 1925, 1941); it involves electronic amplification and a phase-sensitive motor to drive the Variac. According to Schairer (1959), temperatures may be maintained for many months to within ± 0.3 °C with this equipment.

The second method of temperature control utilizes a control thermocouple, attached to (or placed close to) the furnace windings. This control thermocouple is connected to a precision galvanometer which measures the temperature and initiates control by turning the current on or off when the measured temperature deviates from a set value. Most modern controllers are designed to proportion the current to a saturable reactor or other power-amplifying device when temperature deviates from a set point. Using this proportional method, temperatures may be controlled to within 0·25 per cent of the set value. The proportional method also avoids problems of temperature 'overshoot' which is common in the simple 'on'–'off' controllers.

Temperature controllers are discussed further in Chapters 5 and 6.

Temperature calibration in quench furnaces

Until recently very little quantitative data were available on the problems of temperature calibration and control in quench furnaces. For example, Schairer (1959, p. 126), in his description of the quench furnaces and the quenching technique used at the Geophysical Laboratory, states that 'the effect of inhomogeneities in the thermocouple wire away from the junction itself thus become relatively insignificant' and, 'thermal gradients and sample size are such for quenching furnaces that no portion of the sample may vary more than a few tenths of a degree from that of the thermocouple junction, and the method of calibration corrects for a portion of this small difference in temperature'.

Recently, Biggar and O'Hara (1969) have made extensive investigations of temperature errors in quenching furnaces; they conclude that systematic temperature errors as large as 30 °C are possible. In their study, a more elaborate quenching furnace and method of temperature control was used than these described above; the arrangement also differed from that used at the Geophysical Laboratory (Schairer 1959), but the souces of error are the same in both arrangements. Biggar and O'Hara report that at 1000 °C the circumferential and longitudinal temperature gradients gave ranges of 12 °C around the inner furnace wall, and 4 °C per cm in a section perpendicular to the wall. Biggar and O'Hara suspended metal sample-capsules of length 1·5 cm in their furnace and found that the maximum temperature-accuracy was no better than ± 15 °C. They also found that, after adjustment of a set furnace-temperature (using a 1-kW saturable reactor-type controller), several hours were required to re-establish thermal equilibrium, because of 'overshoot' or 'undershoot' of the desired temperature. For experiments with samples of different thermal capacity and reflectivity, temperature differences up to 20 °C were observed. Similarly, thermocouples of different thermal capacity and reflectivity produced large temperature differences, which were functions both of temperature and depth in the furnace tube. Biggar and O'Hara recommend that in individual furnaces (even of the same design) the characteristics of each should be determined independently.

The results of these experiments indicate that great care must be taken in the calibration and control of temperature in quenching furnaces, and that the temperature errors were much larger than previously believed.

Sample preparation and procedure

The preparation of samples for quenching experiments and the techniques employed can be conveniently divided into those involving silicates and those involving sulphides.

Silicates

For equilibrium studies of silicates, the finely crushed sample (as a glass or partly or wholly crystalline material) is placed in a clean platinum-foil envelope, approximately 0·01 mm in thickness and 10 × 16 mm in area. The envelope is made by folding the 10-mm edges together with small tweezers and then folding the 8-mm sides twice, producing an envelope approximately 6 × 8 mm; it is capable of holding about 10 mg of the charge. The top of the envelope is punched with a pin, and a fine platinum wire is inserted to suspend the charge over the hooks of the quenching rig (Fig. 4.4(b)) in such a way that it lies as close as possible to the thermocouple tip. Because platinum tends to weld together at high temperatures, a thin refractory ring is placed between the suspension wire and envelope. A similar arrangement is used if two or more charges are run together. In this case, the charges are separated by refractory rings or filter paper which has been dipped in a suspension of MgO.

The charge is placed on the quenching rig and put into the preheated furnace. To ensure that the charge is properly centred and is not touching the thermocouple tip, the plug is removed from the bottom of the furnace tube and the charge is examined through a small mirror. At temperatures above 1400 °C, this should be done with tinted glass. The plug is then replaced, and the charge is held at the exact temperature for the desired length of time. At the end of this period the plug is removed, and the charge is immediately dropped into a basin of mercury, by passing a current through the suspension wires thus 'shorting' the fine platinum wire which holds the charge. The voltage used in this procedure must be sufficiently low to avoid 'shorting' the wire with explosive force, which might spatter platinum on the thermocouple tip or furnace walls.

The platinum is peeled from the charge with tweezers, and the charge is crushed in a small agate mortar. For initial optical examination, only coarse crushing is necessary; this simplifies the detection of glass. For X-ray determination, the charge should be finely ground. For charges containing appreciable amounts of glass, the platinum foil adheres to the glass, and separate envelopes must be used for each run. In sub-solidus experiments the platinum envelope may be cleaned in hydrofluoric acid and reused.

Sulphides

The preparation of samples and techniques employed in experiments involving sulphides differ from those of silicates, because many sulphide phases cannot be preserved by quenching. Basically, the equipment used is similar; details of experimental techniques are given by Kullerud (1970, 1971).

Starting-materials for most synthetic sulphide studies are usually in the form of the elements themselves. However, sluggish reactions, particularly in the generally low-temperature sub-solidus regions of sulphide systems, often make it necessary to presynthesize some intermediate compound in the elemental binary system and use this compound as starting-material. For the same reason, it is usually not possible to obtain an equilibrium result with a single run, because sulphide rims form on the grains being synthesized; this is avoided by removing the charge from the furnace, regrinding it, and reheating.

The correct proportions of starting-materials are placed in a fused high-purity silica tube,† which has been sealed by fire-polishing in the flame of an acetylene torch. These tubes are approximately 3–7 mm in diameter, with a wall thickness of 0·15–2 mm. Tubes of suitable diameter should withstand saturated sulphur vapour-pressure up to 1000 °C. For example, in synthesizing sphalerite (ZnS) at 850 °C the vapour-pressure of sulphur is 40 atm and of zinc less than 1 atm (Kullerud 1952). A tightly-fitting fused silica plug is inserted in the remaining volume of the tube. The plug reduces the volume, keeps the charge in place, and protects the starting-material from being volatilized on heating and from being sucked out when the tube is evacuated. A wet cloth is wrapped round the lower portion of the tube to prevent oxidation and heating of the starting-materials during sealing. The tube is then connected to a vacuum-pump fitted with a stopcock, and the air slowly pumped out. When a vacuum of 0·02 mmHg is attained, the tube is sealed by rotating it rapidly in a gas flame while under vacuum. The melted silica at the end of the tube collapses on to the silica plug, producing a tight seal. Details of other types of sample tubes for specialized applications are given by Kullerud (1971).

† Below 1100 °C, silica is the only suitable non-reactive material for sulphide reactions, since it does not react with sulphur.

Because of the low melting-point of sulphur, there is no necessity to premix the components of the starting-materials; homogenization may be achieved by gently heating the tube, thus causing the sulphur to melt and produce an intimate mixture.

The charges are placed in the 'hot spot' of a quenching furnace operated in the horizontal position. At the end of the run, the charges are scooped out of the furnace and quenched in cold water. The tubes are broken open, and the products examined by reflected-light optical, X-ray diffraction, or electron-microprobe methods. The quenched products of many sulphide systems must be examined as soon as possible after quenching, before they oxidize or are altered in some other way. Details of the preparation of polished sections for reflected-light microscopy, etc., are discussed by Kullerud (1971).

Studies of sulphide systems can also be made at confining pressures higher than those of the system itself using cold-seal pressure-vessels as described in Chapter 5. For these experiments, samples are placed in sealed gold capsules. The method is discussed by Kullerud and Yoder (1959) and Kullerud (1971). Differential thermal analysis and X-ray heating camera methods are also widely used in investigations of sulphide systems.

Experiments under inert and controlled atmosphere

The importance of certain elements with more than one valency state (particularly iron) often makes it necessary to design and conduct experiments in which the valency of the element is controlled. Under 'dry' conditions this can be done by using an inert atmosphere (such as nitrogen or argon) to prevent oxidation; it can be done in a more quantitative manner by mixing suitable gases to give a controlled atmosphere, producing the desired partial gas-pressure. Under hydrothermal conditions, partial gas-pressures are controlled, using the solid-buffer techniques described in Chapter 8.

If exact knowledge of the partial gas-pressure is unnecessary, the experiment may be done in a suitable atmosphere, using a singlegas. For example, if a sufficiently reducing atmosphere is required to prevent oxidation of magnetite to hematite, nitrogen, argon, or some other suitable gas is passed through the quenching furnace for the duration of the experiment. The gas used must not react with the sample, furnace tube, windings, etc., and must not produce an explosive mixture. The vertical quenching furnace (Fig. 4.4(a)) may be adapted for this type of experiment by fitting both ends of the furnace tube with gas-tight seals (Biggar and O'Hara 1969). The inlet tube is fitted to a tank of commercial gas, and the gas flux is controlled by a micrometer device. Using argon gas, a flux of 2–3 ft^{3}h^{-1} will prevent the oxidation of magnetite to hematite at temperatures up to approximately 1200 °C. The experiments are carried out in exactly the same manner as described above, except for quenching. This is done by raising the quenching rig to the top of the furnace, where water coils enable the sample to be quenched while still in the presence of the inert gas.

A much more accurate method of controlling gas-pressures is to use constant gas mixtures, which are premixed and analysed before passing into the furnace. From the known mixture of the gases, the percentage of oxygen, or other gas, in the system can be determined for different temperatures. Many different combinations of gases can be used; some common ones for controlling $p(O_2)$ are CO_2–CO, CO_2–H, H_2O–H_2, CO_2–CO–H_2O, and CO–O_2.

The gas-mixing apparatus, as described by Darken and Gurry (1945), is shown schematically in Fig. 4.5. The purified gases are passed through two flowmeters F_1 and F_2 consisting of capillary tubes C and manometers G. Darken and Gurry (1945, p. 1399) recommend tube lengths in F_1 and F_2 of 23 cm and 300 cm, respectively. The drop in pressure in each flowmeter is kept constant by bleeding through the outer tubes B, held at fixed levels in a water tank W. The gases are then mixed in tube M which contains glass beads. Some of the mixed gas escapes through the central bleeder tube B; the remainder passes into the furnace at a rate measured by flowmeter F_3. For CO_2–CO and CO_2–H_2 mixtures, Darken and Gurry recommend a linear flow rate of 0·9 cm^{-1}. Using a CO_2–CO mixture, Darken and Gurry found that analyses of the gas before passing into the furnace and after passing out of the furnace differed by only a few tenths of one per cent (up to a ratio of 4:1, CO_2:CO) from the mixture calculated from the calibrated gauges. Details of the gauge calibration and the method for analysing the gas mixtures are given by Darken and Gurry.

The design of the furnace is similar to that for regular quenching experiments, except for the necessary gas-tight seals. Darken and Gurry (1945) give one type of furnace design used for their studies in the iron–oxygen system. More recently, Biggar and O'Hara (1969) have described a more elaborate furnace for regular quenching experiments in air, or for mixed-gas atmosphere work.

The procedure used in experiments with gas mixtures depends on the nature of the investigation. Again, for iron–oxygen investigations, the procedure of Darken and Gurry (1945) is in common use.

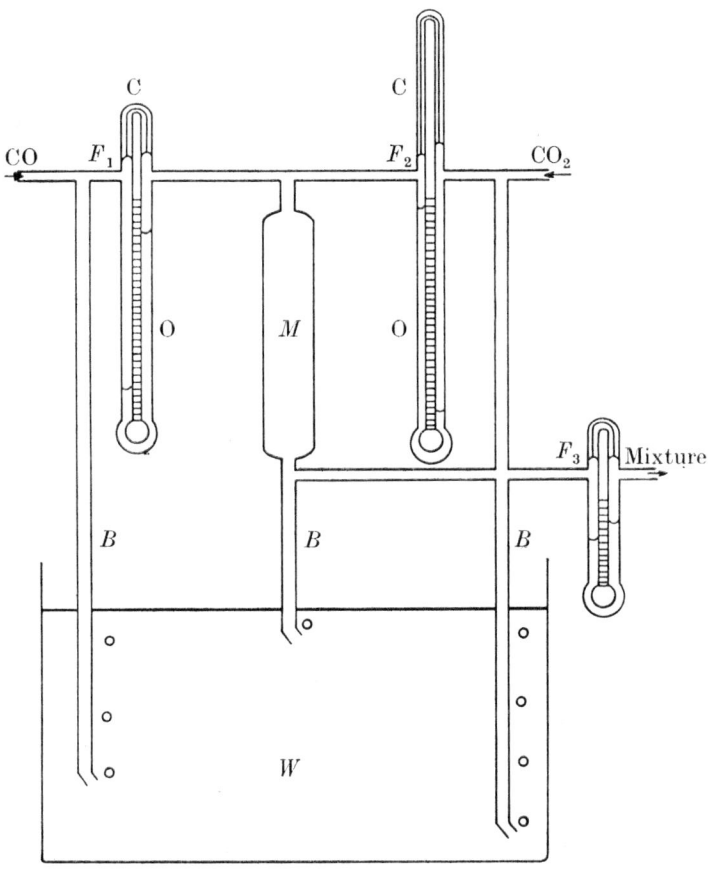

FIG. 4.5. Gas-mixing equipment for controlled atmosphere experiments. (After Darken and Gurry 1945.)

Control of oxygen fugacity by sensing cells

The determination of oxygen fugacity, using the gas-mixture method of Darken and Gurry (1945), involves careful calibration of both the flowmeters (through which the gas passes) and the final gas mixtures. Record (1970) and Sato (1970, 1971) have described high-temperature oxygen-concentration cells (or oxygen-probes) which eliminate final gas analysis and either simplify or eliminate calibration of flowmeters. This method measures oxygen fugacities, directly without removing the sample from the furnace; it has an extremely rapid response-rate. The probes can be used both for the measurement of very low oxygen fugacities and for a wide range of oxygen concentrations.

The oxygen-probe consists of a cell containing two platinum electrodes, which are in physical contact with a gas-tight solid electrolyte capable of electrical conduction solely by oxygen ions. The general characteristics of solid electrolytes are discussed by Sato (1971). He recommends lime-stabilized zirconia tubes (15 mol per cent CaO), or suggests yttria-stabilized zirconia tubes (10 mol per cent Y_2O_3) as solid electrolytes. With this arrangement, a potential difference (e.m.f.) develops between the two electrodes, which depends on the oxygen fugacities at either side of the electrolyte according to the equation

$$E = 2\cdot 303 \frac{RT}{4F} \log_{10}[p(O_2)/p(\text{ref})] \qquad (4.3)$$

or

$$E = 0\cdot 0496 T \log_{10}[p(O_2)/p(\text{ref})] \qquad (4.4)$$

where E = e.m.f. in volts eqn (4.3) or mV eqn (4.4),
R = gas constant ($8\cdot 3143$ JK^{-1} mol^{-1}),
T = absolute temperature,
F = Faraday constant ($9\cdot 6487 \times 10^4$ C mol^{-1}),

$p(O_2)$ and $p(\text{ref})$ = partial pressures of oxygen at the two electrodes. If a suitable reference gas of known composition, such as air,† with a constant oxygen content of 20·9 per cent by volume is pumped into one electrode, the partial pressure of oxygen at the other electrode can be determined for the potential-difference.

The temperature range in which the oxygen-probe can be accurately used is governed by the amount of electronic conductance (as opposed to pure-oxygen ionic conductance) in the solid electrolyte and by its purity. For lime-stabilized zirconia, the useful range lies between 600 °C and 1150 °C at $p(O_2)$ between 10^{-33} and 10^{-17} atm. Measurement of $p(O_2)$ requires an allowance for the temperature of the probe output, measurement of oxygen potential μO_2 does not require this allowance.

A sketch of the oxygen-probe is given in Fig. 4.6. The stabilized-zirconia tube is closed at one end and serves as the electrolyte. Platinum foil (approximately 0·0005 in thick) is attached to the inner and outer surfaces of the closed end of the zirconia tube by platinum paste or platinum wire to form the two electrodes. Platinum wires are attached to the foil and connected to the measuring device. Alumina tubes, one

† Air cannot be used if contamination from the environment is likely.

inside the zirconia tube and the other outside, protect the electrodes; the probe itself is protected by a suitable sheath (not shown in Fig. 4.6) of some heat-resistant material, such as silicon carbide. Sato (1970) gives a drawing of a more elaborate arrangement which includes a thermocouple for temperature measurement. The oxygen-probe itself may be used as a thermocouple by taking the platinum lead as one wire of the

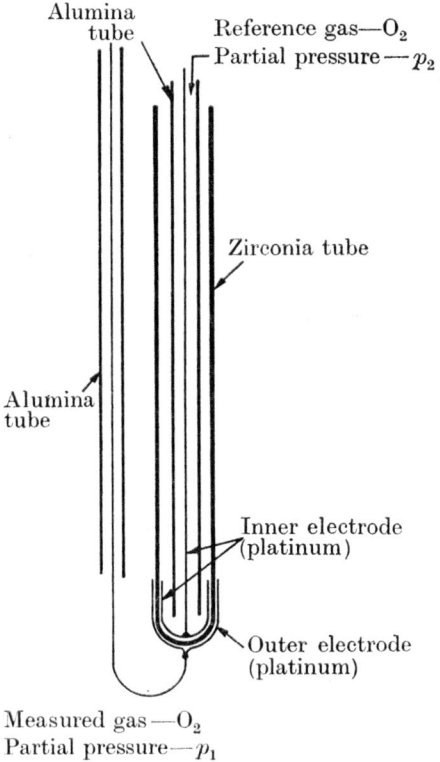

FIG. 4.6. Simple oxygen probe. (Modified after Record 1970.)

couple and inserting a separate platinum–rhondium lead. With this arrangement the values of $p(O_2)$ (as e.m.f.) and T (as e.m.f) may be monitored alternately with a single measuring device. Alternatively, a four-bore alumina insulator may be placed inside the electrolyte tube. Two bores contain the thermocouple wires; one the platinum electrode lead, and the fourth allows access of the reference gas. Either method is more advantageous than the conventional type of thermocouple, since it measures temperature at the probe tip, thus recording any cooling arising from an excess of reference gas passing through the probe. If such cooling is not detected, erroneous values of $p(O_2)$ will result (eqns 4.3 and 4.4).

The reference gas is pumped into the tube, which is slowly inserted† into the hot zone of the gas-tight furnace. Mixtures of appropriate gases are passed into the furnace through a flow-regulating device, and the mixing-ratio of the gases is adjusted (using the previously described method of Darken and Gurry 1945) until the e.m.f. of the probe equals the e.m.f., calculated from eqn 4.4, necessary to produce the

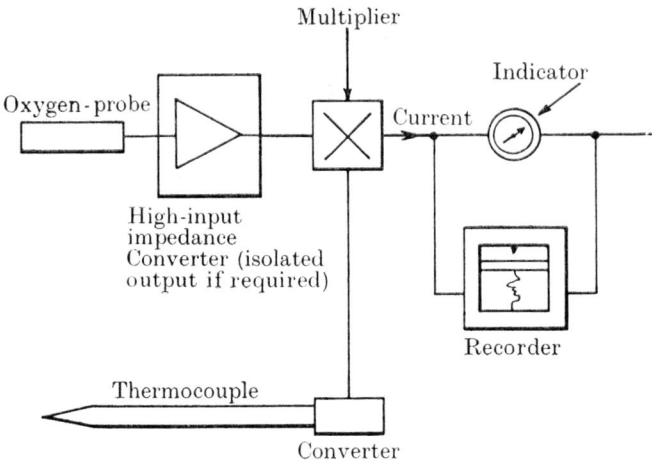

Fig. 4.7. Simple measuring system for measurement of oxygen potential concentration (modified after Record 1970.)

desired oxygen fugactiy. Sato (1970) describes an electrochemical method for regulating the oxygen fugacity in the furnace. Hydrogen is generated by electrolysis of water, and then mixed with water vapour and other gases. This method eliminates the necessity for calibrating the flowmeters.

A sketch of one possible measuring system for oxygen potential or concentration is shown in Fig. 4.7. It consists of a high-input impedance converter (capable of measuring probe output under minimum operating temperatures, when source impedance is highest), a standard multiplier, an indicator, and (if necessary) a recorder. For measurements of oxygen potential $\mu(O_2)$, a temperature-measuring system may be added in the form of a thermocouple system. Sato (1971, Fig. 13) gives a sketch of a

† Record (1970) recommends that the probe temperature should not increase by more than 100 °C min^{-1}.

system incorporating a method which minimizes contamination of the sample by diffusion of gas through the electrolyte tube.

In a vertically operated gas-tight quenching furnace (Biggar and O'Hara 1969), the oxygen-probe is inserted into the lower end of the tube and the quenching rig into the upper end. The sample, recording thermocouple (if not incorporated in the probe itself), and oxygen-probe tip are as close together as possible. At the end of the experiment the sample can be quenched by raising the quenching rig to the top of the furnace tube, around which water is circulated. In this way, the sample is quenched while the gas atmosphere is maintained.

Other possible sensing cells of petrological interest are the sulphur, fluorine, and hydrogen cells. Sato (1971) briefly discusses these.

References

BIGGAR, G. M. and O'HARA, M. J. (1969). *Miner. Mag.* **37**, 1.
DARKEN, L. S. and GURRY, R. W. (1945). *J. Am. Chem. Soc.* **67**, 1398.
FAUST, G. T. (1936). *Am. Miner.* **21**, 735.
FORSYTHE, W. E. (1941). in *Temperature. Its measurement and control in science and industry.* Reinhold, New York. p. 1115.
KULLERUD, G. (1952). *Norsk geol. Tiddskr.* **32**, 61.
—— (1970). *Miner. Soc. Am. Special Publication No. 3*, p. 199.
—— (1971). in *Research techniques for high pressure and high temperature* (ed. G. C. Ulmer). Springer–Verlag, New York. p. 289.
—— and YODER, H. S. (1959). *Econ. Geol.* **54**, 53.
MARGRAVE, J. L. (1959). *Physiochemical measurements at high temperatures* (ed. J. O. Bockris) Butterworths, London. p. 6.
RECORD, R. G. H. (1970). *Instrum. Pract.* **24**, 161.
ROBERTS, H. S. (1925). *J. opt. Soc. Am.* **11**, 171.
—— (1941). in *Temperature. Its measurement and control in science and industry.* Reinhold, New York. p. 604.
ROESER, W. F. (1941). in *Temperature. Its measurement and control in science and industry.* Reinhold, New York. p. 180.
—— and WENSEL, H. T. (1935). *J. Res. Nat. Bur. Stand.* **14**, 247.
SATO, M. (1970). *Am. Miner.*, **55**, 1424.
—— (1971). in *Research techniques for high pressure and high temperature* (ed. G. C. Ulmer). Springer–Verlag, New York. p. 43.
SCHAIRER, J. F. (1959). in *Physiochemical measurements at high temperatures* (ed. J. O. Bockris) Butterworths, London. p. 117.
SHEPHERD, E. S., RANKIN, G. H., and WRIGHT, F. E. (1909). *Am. J. Sci.* **28**, 293.
SOSMAN, R. B. (1952). *Am. J. Sci.* (Bowen Vol.) p. 517.

5. Externally-heated pressure-vessels

Introduction

EXTERNALLY-heated pressure-vessels, consisting basically of a metal alloy test-tube with a pressure seal, can be used up to about 10 kbar gas pressure (commonly H_2O) at temperatures of 750 °C, and at much higher temperatures under lower pressures. Although pressure-vessels have been in use for almost a century, it is only within the last twenty-five years or so that they have been extensively used for petrological and mineralogical research. Before the designs of Morey (1918) and Morey and Fenner (1917), vessels consisted of tubes loaded with solids and water and sealed, either by welding both ends of the tube or by means of a simple flat washer and screw. Obviously such vessels were difficult to maintain and dangerous to operate.

In the Morey design, discussed in the following section, a much superior sealing arrangement was used. With the development of the Tuttle vessels (Tuttle 1948, 1949) and their subsequent modifications for higher pressure and temperatures, the use of externally-heated pressure-vessels became routine for phase-equilibrium studies, and they have probably provided more information of geological interest than any other single piece of equipment (Wyllie 1966).

Types of pressure-vessels

Morey vessels

In the original Morey vessels (Morey and Fenner 1917; Morey 1918), both the vessel and closure consisted of tool steel which could withstand pressures of 400 bars at temperatures up to 600 °C for long periods or 100 bars at 700 °C for short periods. The chief improvement in this design over earlier ones was in the closure arrangement (Fig. 5.1(a)). This consisted of a thin copper or silver washer which was compressed as the threaded closure nut was tightened. Morey and Ingerson (1937) described a smaller version using stainless steel for the vessel and tool steel for the closure.

Two difficulties are inherent in this design. First, the thrust of the closure nut on the soft metal sealing gasket produced leaks at high

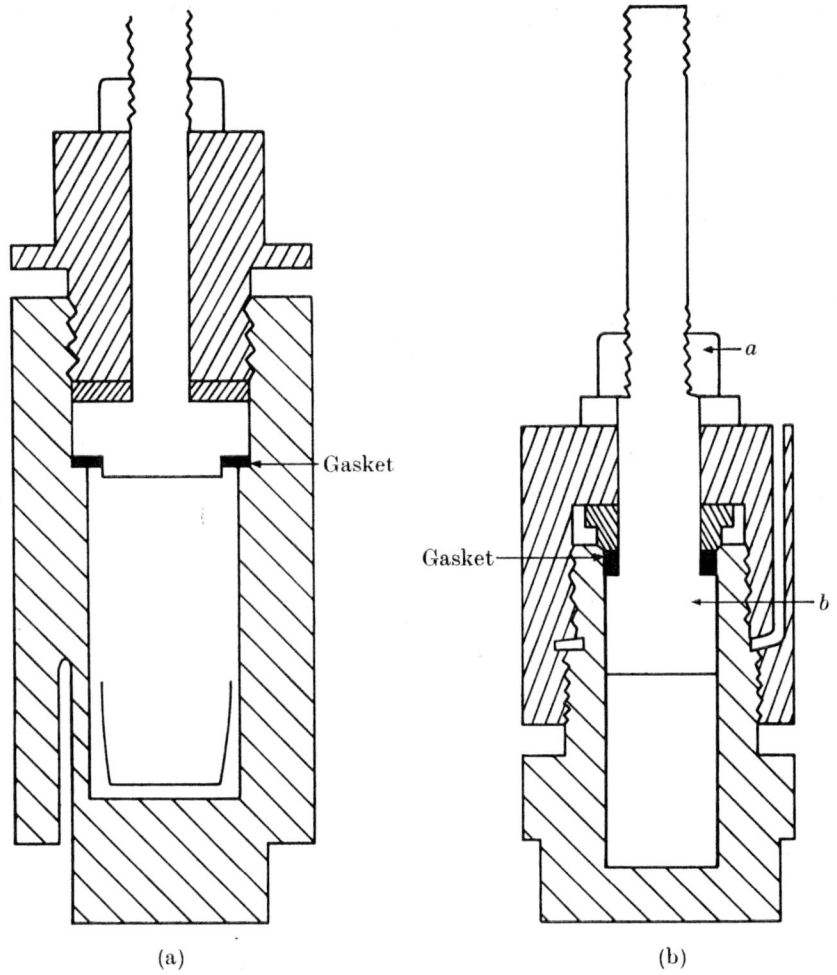

FIG. 5.1 Morey-type pressure-vessels: (a) Design of Morey and Fenner (1917) and Morey (1918); (b) Design of Morey (1953).

pressures or with rapid temperature changes. This made operation of these vessels time-consuming, not only because of leaks but also because of difficulties in removing the closure at the end of the run. The second problem with these vessels was in pressure measurements, which involved calculation of the pressure from P–V–T relations, by sealing a known volume of water into the vessel at the beginning of the experiment and assuming that none was lost by leakage during the run.

These problems were alleviated in a later design (Morey 1953,) shown in Fig. 5.1(b), in which closure is made by a Bridgman unsupported area seal gasket made of copper or silver. Pressures as high as 4 kbar at 600 °C may be attained with an Inconel vessel of this type (Morey and Hesselgesser 1951, 1952; Morey 1953). At low pressures sealing is achieved by compression when nut a (Fig. 5.1(b)) is tightened. With increasing pressure, this seal becomes tighter as the pressure within the vessel is applied to the head of seal b, producing a pressure on this gasket about twice that inside the vessel. The pressure can be directly measured and adjusted during an experiment by providing an axial hole through the closure nut (not shown in Fig. 5.1(a)).

Because of their large volume, Morey vessels are useful for experiments with large samples. These are placed in suitable containers and exposed to the pressure medium. In this respect Morey vessels differ from later designs in which the sample is separated from the external pressure medium. For some types of experiment this may be a disadvantage, for the pressure medium may selectively remove material from the hotter portions of the vessel and deposit it in the cooler portions, with the result that the sample changes composition during the experiment.

The general layout for the pressure and temperature supply system, pressure piping and control, and temperature control and measurement is given in a later section (p. 100 *et seq.*).

Weighed amounts of the sample are placed in a suitable container, usually a platinum crucible or capsule, and distilled water is added. In the early designs, the volume of the vessel was accurately known and the pressure within the vessel should be calculated from the exact volume of water added. With later designs which have an axial hole though the closure this is unnecessary.

The closure is then tightened, a thermocouple is inserted in the well close to the sample, and the vessel placed inside a suitable furnace. The entire Morey vessel and closure lie within the heating element of the furnace; this is in contrast to cold-seal designs in which the closure is placed outside the furnace. At the end of the run, the vessel is quenched in a jet of air or in water and the closure seal broken. In many of the early designs, removal of the closure was often difficult, but Kennedy (1950) has shown that this is easier if the threads are tapered.

Morey vessels may also be used to investigate systems with two volatile phases, such as H_2O and CO_2. Details are given by Greenwood (1962, 1967*a*, *b*) and Gordon and Greenwood (1970, 1971). Briefly, this technique involves placing one of the gases in the cold vessel. When the desired temperature is reached, the first gas is adjusted and the second one added until the required composition and pressure of the fluid is attained. At the end of the run, the fluid is extracted from the vessel and analysed. Gordon and Greenwood (1971) have also adapted a method initially proposed by Holloway et al. (1968) in which partial pressures of CO_2, H_2O, and hydrogen are generated by decomposition of oxalic acid and water in a large crimped-platinum capsule inside which is a smaller platinum capsule, also crimped, containing the desired starting-material. Under the conditions of the experiments, the water-pressure medium has an oxygen potential approximately equivalent to that of the Ni–NiO buffer (see Fig. 8.1). This allows excess hydrogen from the decomposition of the oxalic acid to diffuse out of the capsule. Using Morey vessels and these techniques, gas compositions are accurate to within ± 0.3 per cent using gas mixtures (Greenwood 1967a) and to ± 1 per cent using oxalic acid (Gordon and Greenwood 1971).

Tuttle-type vessels

A pressure vessel of much smaller volume, which is both cheaper and easier to use, was designed by Tuttle (1948). This vessel, shown in Fig. 5.2(a), consists of an unthreaded stainless-steel or Stellite 25 (cobalt-alloy) 'test-tube', closed by a cone-in-cone seal. As with the Morey vessels, the entire vessel and seal are within the furnace. The seal is closed by weights acting through heat-resistant alloys above and below the vessel. The cone at the lower end of the support rod is machined to 59°, the cone on the vessel to 60°; this results in a ring seal. At pressures in excess of 2 kbar and above 800 °C, these cones have to be machined after each experiment.

Pressure is transmitted to the sample through an axial hole in the lower support rod, and temperature is measured by placing a thermocouple in a well of the vessel within 3 mm of the sample. A vertical split furnace is mounted on a hinge, allowing it to be swung clear of the vessel during loading or quenching. This vessel can be used up to 800 °C at 4 kbar $p(H_2O)$. In contrast to the Morey vessels, experiments in this apparatus are done in a 'closed' system in which the sample and water are sealed in a noble metal capsule using the method of Goranson (1931), which is described later in this chapter.

In 1949, Tuttle described a modification of his earlier design which is much easier to use. This consisted of a longer vessel in which the open end and seal are outside the furnace (Fig. 5.2(b)), hence the term 'cold-seal vessel' (although in actual fact the seal is far from cold). An advantage of this design is that the cone-in-cone seal is held in place by a right-hand threaded nut cap. When this cap is tightened, pressure is

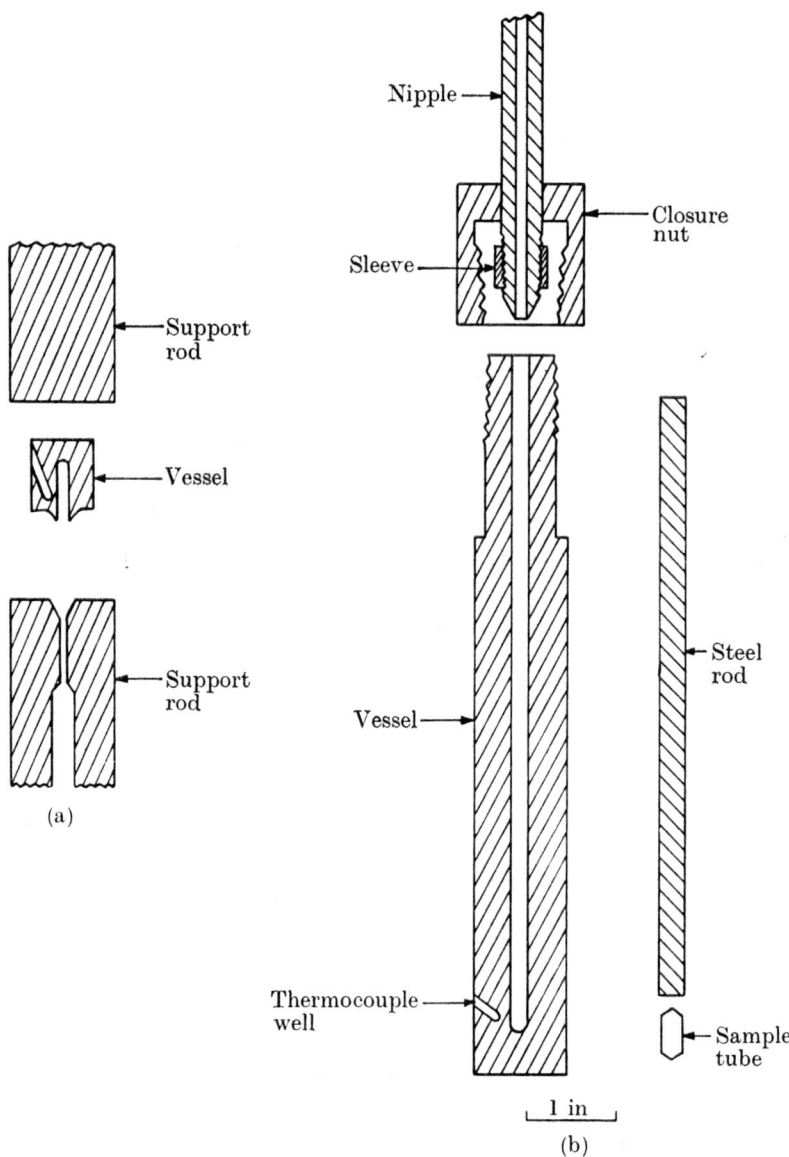

FIG. 5.2. Tuttle-type pressure vessels: (a) Design of Tuttle (1948); (b) Design of Tuttle (1949).

exerted on a left-hand threaded sleeve on the end of the nipple on the closure stem (Fig. 5.2(b)), thus eliminating the need for weights to keep the seal tight. In this design, the closure can be made of stainless steel since it remains relatively cool. Pressure is transmitted to the sample, which is contained in a sealed capsule, through a hole in the closure. Temperature is measured either by placing a thermocouple in a well of $\frac{1}{16}$ in diameter in the side or base of the vessel as close as possible to the sample, or by using an internal, Inconel-sheathed thermocouple which passes through the closure and is sealed by a steel cone silver-soldered on to the sheath (Rooymans 1967).

Tuttle vessels of this type may be operated closure-up, closure-down, or horizontally. Whichever position is used there is a tendency for large thermal convection currents to develop because of the temperature differences between the fairly cool closure and the sample capsule in the 'hot spot' of the furnace. These may be reduced by placing a back-up or filler rod of corrosion-resistant material between sample capsule and closure. This rod also reduces the danger of explosion by minimizing the volume of the pressure medium, and keeps the sample in the 'hot-spot' of the furnace.

The ratio of the vessel diameter to hole diameter (wall diameter) determines the strength of the vessel. In most standard Tuttle vessels this is about 4:1 for a vessel 8–12 in long. Using Stellite 25, operation to about 900 °C at 1 kbar $p(H_2O)$ or 750 °C at 3 kbar $p(H_2O)$ is possible for long-term runs.

Since Tuttle's original designs in 1948 and 1949, a number of improvements in closure design have been made using new alloys which have extended the $p(H_2O)-T$ range of cold-seal vessels. Some of these designs are described later in this chapter.

Luth and Tuttle (1963) describe a modified Tuttle-type closure, as shown in Fig. 5.3, which permits vessels 8 in long, $1\frac{1}{4}$ in outside diameter, and $\frac{1}{4}$ in inside diameter to be used up to 10 kbar $p(H_2O)$ at temperatures as high as 750 °C. Rene 41 (a nickel-based alloy) is used for this vessel. In this design a back-up washer is placed on a hardened cone; this prevents rotation of the cone when the closure nut is tightened.

Fawcett et al. (1971) claimed that Luth and Tuttles' closure could not be used repeatedly at high pressure without remachining the parts, and they proposed the closure shown in Fig. 5.4. In this design, the cone is thrust into the vessel and tightened by three high-tensile bolts through the cap nut on to the thrust washer at the back of the closure cone. In order to maintain perfect alignment, a guide piece for the cone threads directly into the vessel. Careful matching of the diameters, and only hand tightening of the cap, prevents slow leaks at high pressure even in long experiments.

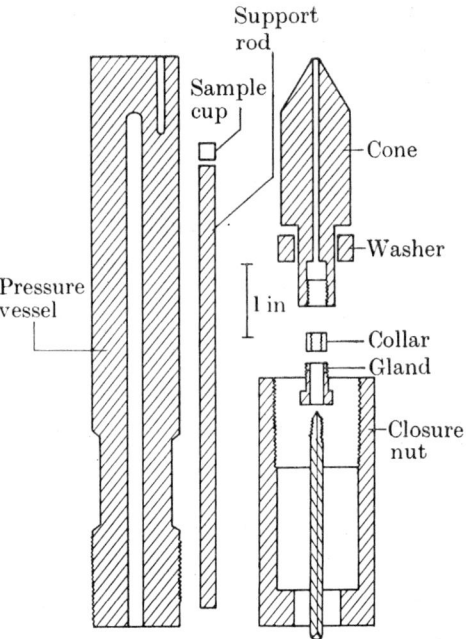

FIG. 5.3. Luth and Tuttle-type pressure-vessel. (After Luth and Tuttle 1963.)

Althaus (1969) describes vessels with a perfectly hemispherical closure and wall ratios up to 7:1 for experiments at 750 °C and 12 kbar $p(H_2O)$. In all experiments in excess of about 7 kbar, argon is used as the pressure medium, since water freezes at room temperature at these pressures.

Operation of Tuttle-type vessels

In this section suggestions are made for the design and operation of hydrothermal† experiments using cold-seal pressure vessels. Readers may wish to make their own modifications, and only the basic components for pressure and temperature generation and operating techniques are described here.

Basically, hydrothermal quenching experiments do not differ greatly from the quenching experiments at atmospheric pressure described in

† See footnote on p. 7.

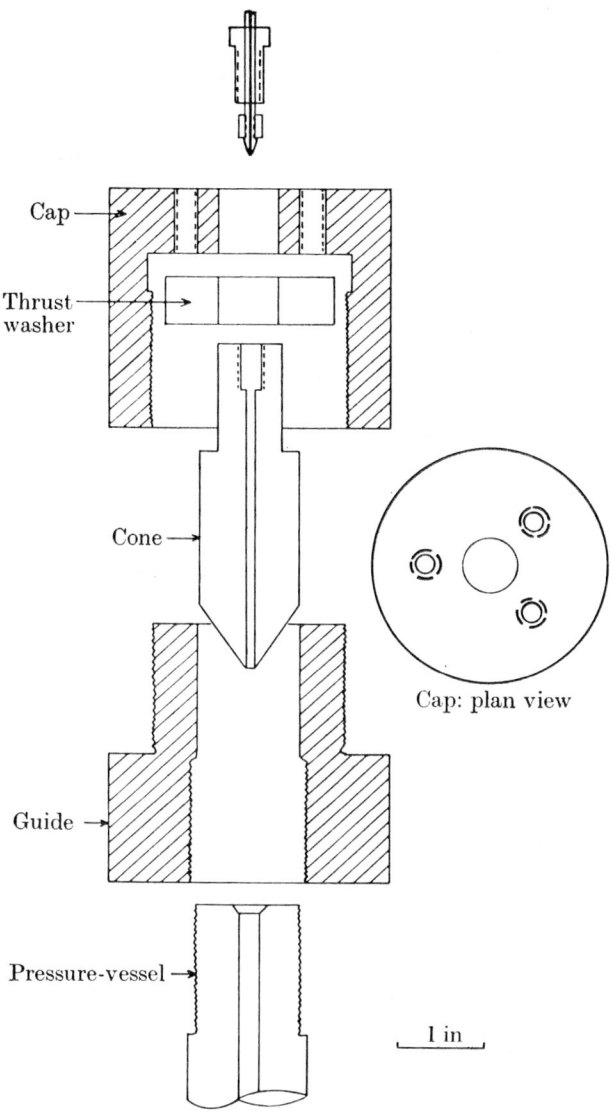

FIG. 5.4. Fawcett *et al.*-type pressure-vessel. (After Fawcett *et al.* 1971.)

Chapter 4. However, the added pressure variable makes this technique slightly more difficult and considerably more dangerous. Some precautions against the commoner hazards are described later in this chapter. A schematic diagram of typical hydrothermal quenching equipment is shown in Fig. 5.5.

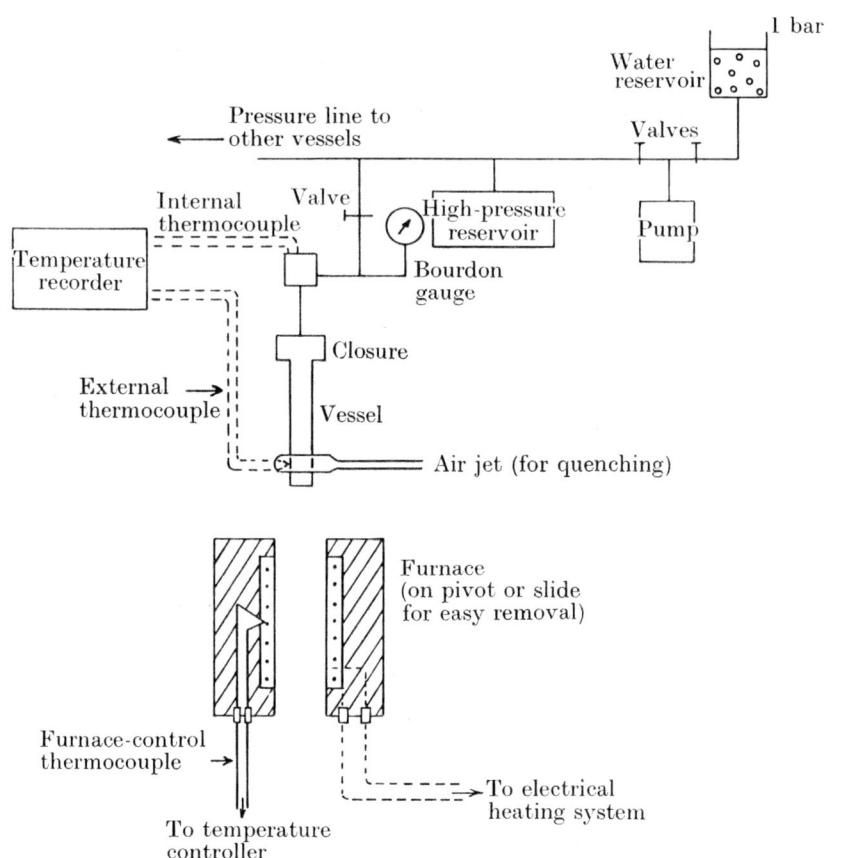

FIG. 5.5. Schematic pressure–temperature arrangement for externally heated pressure-vessel.

Pressure system and accessories

The pressure system shown in Fig. 5.5 is for operation up to 2–3 kbar. The pressure medium is usually distilled water; a little glycol may be added to inhibit corrosion. For higher fluid pressures, an hydraulic intensifier or air-driven pump is used, with argon or nitrogen as the pressure medium. These are described in Chapter 6.

Externally heated pressure-vessels

The components of the pressure system shown in Fig. 5.5 are low- and high-pressure reservoirs, a pump, various valves, and a pressure gauge. The pump may be of the manual type† in which water is forced through an aperture of suitable size by means of a hand-operated piston. The pump is connected to the 1-bar reservoir by a non-return valve. Air-driven pumps‡ are also very convenient for this type of work since they operate from a laboratory, or bottled, air supply and can be regulated to deliver a constant pressure. Usually less than 100 lb in^{-2} is required to produce a water pressure over 2 kbar. This type of pump is particularly helpful in quenching, which is normally done isobarically.

If a hand pump is used, a high-pressure reservoir is incorporated in the pressure line. This reservoir helps to maintain pressure during quenching and to prevent excessive rise in pressure during heating. A valve separates the pump from the high-pressure reservoir.

Normally several vessels are run from one pump, each vessel being connected to the high-pressure line by a valve and T-junction. The valve (two-way type) isolates the vessel from the line. Pressures in the vessel are measured by a Bourdon-tube type gauge with a safety-glass dial and 'blow-out' back. For high pressures, manganin cells are used to measure pressure, as discussed in Chapter 6. If many experiments are being done at the same pressure, only one gauge is required. Otherwise each vessel should have a separate gauge.

The types of pressure tubing, valves, and fittings used depend on the operating pressures. For the main pressure lines, stainless-steel or steel alloy tubing in a variety of sizes, rated for pressures of 2 to about 13 kbar is available. The minimum size recommended is $\frac{1}{4}$ in o.d., $\frac{1}{16}$ in. i.d. Connections between pressure tubing and other fittings are made by a 59° conical seat on the end of the tube which fits into a 60° seat on the fitting. The tube has a left-hand thread and sleeve. When a gland nut, with right-hand thread, is fitted over this sleeve and screwed into the fitting, the sleeve is tightened at the same time as the conical seating surfaces are sealed. Normally this can be done with an ordinary hand wrench to seal pressures up to 13 kbar. It is strongly recommended that T- and L-junctions be used rather than bending pressure tubing. If pressure tubing has to be bent, the radius of bend should be at least six times the diameter of the tube. For pressures below about 4 kbar,

---

† The American Instrument Co., Silver Spring, Maryland, make a suitable pump of this type.

‡ SC Hydraulic Engineering Corporation, Los Angeles, California, make this type of pump.

flexible (or capillary) pressure tubing is useful. This consists of $\frac{1}{16}$-in o.d., $\frac{1}{64}$-in i.d. tubing, which may be connected to the pressure vessel and rigid pressure line by silver-soldering it on to a nipple with a left-hand threaded sleeve as described above. Such an arrangement allows a pressurized vessel to be lifted in and out of the furnace.

Numerous types and sizes of valves, suitable for sealing pressures up to about 4 kbar, are commercially available. The commonest of these are of the needle-valve type, which are closed by hand tightening of the valve stem into the manifold. Such stems may be of the rigid or 'floating' type. The seal is maintained by a series of packings. Once the required pressure is attained, only the high-pressure ports of the valve are pressurized; the packings are not subjected to pressure. Such valves should not be overtightened or forced shut, since this is liable to snap off the stem. Valves for pressures in excess of 4 kbar are described in Chapter 6.

Furnaces and temperature recording and control devices

Furnaces for hydrothermal experiments may be placed vertically with closure-up, as shown in Fig. 5.5; vertically, with closure down; or horizontally. If flexible pressure tubing is used, and the vessels are quenched by removing them from the furnace by hand, split-type furnaces are most useful. If quenching is achieved by raising or lowering the furnace manually or by a remote-control pulley system, a vertical arrangement is desirable.

For cold-seal vessels 8–12 in long and of 1–1$\frac{1}{2}$ in diameter, the furnace should be at least 4 in longer than the vessel and its diameter about $\frac{1}{8}$ in larger than that of the vessel. These dimensions minimize thermal gradients in the vessel and enable an external recording thermocouple to be inserted readily.

Furnaces are normally wound with Nichrome V† or similar base metal wire (see Chapter 4, p. 77), since temperatures in excess of 1200 °C are seldom required in this type of experiment. For non-split tubular furnaces, wires are wound on the outside of a ceramic tube and then embedded in alumina cement, as described in Chapter 4. The control thermocouple is cemented close to the windings in the 'hot spot' of the furnace. If a split furnace is used the element coils are placed in grooves running longitudinally inside two hemispherical ceramic tubes, which can be readily replaced by simply slipping them into the furnace casing. The control thermocouple is cemented close to the 'hot spot', which in a

† Trade mark of the Driver-Barris Corporation.

furnace 12–14 in long should be at least $1\frac{1}{2}$ in wide, or by threading a platinum resistance element on one of the winding coils.

Insulation is provided by powdered MgO (lightly calcined), or suitable insulating bricks cut to size, placed between the ceramic furnace tube and a rolled stainless steel or aluminum sheet. The ends of the furnace are enclosed with Transite or other suitable material, through which the leads for the furnace windings and control thermocouple are attached by insulated terminals.

For most hydrothermal experiments, a control and recording thermocouple are used, as shown in Fig. 5.5. The control thermocouple, close to the furnace windings, is connected to a suitable temperature controller which regulates furnace temperature to ± 2–$3\ °C$ under optimum conditions. Pt–Rh thermocouples are often used for this purpose because of their stability over long periods of time. The much cheaper Cr–Al thermocouples can be used but they must be changed frequently, since they deteriorate rapidly owing to oxidation and other contamination. The recording thermocouple measures the temperature of the sample. This may be either an external thermocouple, inserted in a well on the side or base of the vessel as close as possible to the sample (Fig. 5.5), or an internal, sheathed thermocouple, inserted through the closure of the vessel. Cr–Al thermocouples are generally used for recording thermocouples; external types are replaced after each run.

Construction of thermocouples of the external type is discussed in Chapter 4. Since it is essential that the thermocouple wires do not touch one another or the side of the vessel, insulation is provided by double-bore ceramic tubing, thinner 'spaghetti' tubing being used to insulate the wires in the narrow thermocouple well (Figs. 5.2(b), 5.3). It is important that the thermocouple well should be of the smallest possible diameter ($\frac{1}{16}$ in is standard for a vessel 1 in in diameter) and should not be drilled too close to the main axial hole of the vessel, since stresses in the two holes may result in their convergence.

If internal thermocouples are used, a steel cone is silver-soldered on to the sheath and the thermocouple placed inside the vessel by passing it through a T-junction above the closure (Fig. 5.5). The thermocouple then passes through the main pressure tubing, through the closure, and down through the centre of the filler rod. For this operation, $\frac{1}{4}$-in diameter high-pressure tubing makes a convenient filler rod. An alternative arrangement for very thin sheathed thermocouples (1 mm in diameter) has been described by Rooymans (1967). This entails silver-soldering the thermocouple on to the wall of standard $\frac{1}{4}$-in diameter pressure tubing.

Internal thermocouples have the advantage of recording the sample temperature under pressure. Some workers suggest that temperatures recorded this way may vary by several degrees from temperatures recorded at atmospheric pressure by external thermocouples.

With either type of thermocouple, temperatures are recorded on a potentiometer. If a number of vessels are being operated simultaneously, a multi-point strip chart recording potentiometer, which permits continuous recording of up to 24 different temperatures, is desirable. Such potentiometers require frequent calibration.

The basic principles of temperature controllers have been described in Chapter 4. Most laboratories are now equipped with transistorized temperature controllers. These fully proportional controllers give much better control than the older, non-transistorized types, provided that the temperature–time lag between winding and sensing thermocouple materials is small. Hadidiacos (1969) has described a solid-state instrument of this type which controls to $\pm \frac{1}{2}$ °C and is much less expensive than commerically available models. For hydrothermal experiments it is essential that temperature controllers have a thermocouple-break protection device which automatically sets the controller on a cooling cycle should the control thermocouple break or the controller stick on the 'on' position. This device should be checked regularly, since a malfunctioning controller can overheat a pressurized vessel and cause failure by explosion.

Preparation of samples and run procedures for Tuttle-type vessels

For experiments in cold-seal pressure vessels, the sample is enclosed in a sealed noble-metal capsule together with a volatile phase, usually water. Pressure is transmitted to the sample by collapse of the walls of the capsule when the external presure is applied. When pressure and temperature equilibration are attained, the external pressure is equal to the pressure of the volatile phase within the capsule.

The metal used for the capsules must not react with the sample material; it must be able to withstand the temperatures required; it must be sufficiently malleable to collapse under the external pressure applied; and it must be capable of being easily sealed by welding. The metals most commonly used are gold, platinum, and silver–palladium, in the form of tubes. Gold and platinum are inert to most starting-materials, are malleable, and are easily welded. Gold can be used for experiments up to about 1000 °C, and platinum up to the temperature limits of all cold-seal pressure vessels. Silver is restricted to temperatures below about 925 °C and is much more difficult to seal than gold and platinum. Silver–palladium tubes, often $Ag_{70}Pd_{30}$, are used for samples containing iron in which platinum tubes cannot be used because of alloying of iron and platinum. Some hydrothermal experiments call for samples containing more than one type of capsule. These are discussed in Chapter 8 but the principles of preparation are the same as those for single-capsule techniques described here.

The size of capsule used depends on the type of experiment, the size of the pressure-vessel, and the amount of sample required. For standard quenching experiments, the length of the capsule should not exceed the length of the 'hot' zone of the furnace, and should preferably be less to minimize thermal gradients within the capsule. Gold capsules about $\frac{3}{4}$ in long are suitable for most quenching experiments and can hold approximately 50 mg of an 'average' silicate starting-material and water. If the wall of the tube is too thin, the capsule will tend to burst under pressure. A common wall thickness for gold capsules is $\frac{1}{64}$ in. Because of their greater cost, platinum capsules are usually slightly smaller. Gold and platinum are conveniently purchased in 12-in tubes, which can be cut into the capsules of the desired lengths. One end is then flattened by pliers and welded. After welding, the capsules are then annealed in a bunsen or meker burner.

Two methods of welding are commonly used: gas-welding, using a gas such as acetylene; or d.c. arc-welding, using a carbon electrode. For both mehods, the capsule is held in a narrow-jawed jeweller's vice. The end to be welded is pressed tightly together with a pair of small pliers to ensure that the edges to be welded are perfectly coincident.

For gas-welding, the flame must be regulated to a fine point, which is passed over the pressed end of the capsule as quickly as possible to prevent excessive heating of the capsule, which, if loaded, may then also suffer loss of water. For arc-welding, a pure, well-sharpened carbon-electrode pencil forms one end of the circuit, the other end being attached to the vice containing the capsule. When the carbon electrode touches the capsule the circuit is closed. The electrode is drawn steadily across the flattened end of the tube and the metal melts, forming a smooth weld. With experience, welding can be completed with one stroke of the electrode. If a complete weld is not made in the first attempt, any holes can be sealed by a further application of the electrode. However, prolonged use of the electrode tends to make the capsule excessively hot, resulting in water loss or even explosion of the capsule owing to build-up of vapour pressure. It is essential that the end of the capsule be absolutely clean and dry, since any impurity will make welding almost impossible. Excessive heating of capsules can be avoided if a capacitance is introduced into the d.c. circuit; this permits welding by a series of sparks. With this method a 'clean' weld is more difficult. Whichever method is used, the weld should be examined under a low-powered binocular microscope to check for any holes or other signs of incomplete sealing. Platinum is easier to arc-weld than gold, and silver is difficult owing to its tendency to oxidize. This may be avoided by welding silver tubes in a stream of inert gas such as argon.

Distilled water is added to the weighed sample capsule by a microsyringe, care must be taken not to allow any water to stick to the upper part of the tube, which is wiped dry before reweighing. The weight ratio of water to solid starting-material depends on the type of experiment being done. If $p(H_2O) = p_{total}$, water in excess of that required to saturate the molten solid is necessary. For most silicates, this requires a 1:3 to 1:5 ratio of water to solid starting-material. The exact volume of water to give a required weight may be measured with the microsyringe. In experiments in which $p(H_2O) < p_{total}$,† the ratio of water to starting-material is much more critical and careful measurement must be made. If an excess of water is placed in the capsule it can easily be removed with a tissue or extracted by the microsyringe.

The starting-material is inserted into the capsule with a small spatula‡ and tamped down with a wooden stick or by tapping the capsule on a hard surface. The top of the capsule is thoroughly cleaned with tissue, squeezed shut with pliers, and weighed. For a standard 1-in long gold tube about $\frac{1}{16}$–$\frac{1}{8}$ in should be evenly exposed above the jaws of the vice for convenient welding.

In order to avoid loss of water due to evaporation during welding, the bottom of the tube is wrapped in a wet tissue to keep it cool. Welding in a d.c. arc is also facilitated by cutting a sliver off the top of the capsule using a special pair of small jeweller's pliers. This produces a fresh, even surface and permits easy flow of the metal during welding. The removed material must be retained and included in the final weighing.

Welding a filled capsule is considerably more difficult than an empty one, and material may be lost either by spattering of the metal or by evaporation of the volatile. When welding is complete, and the seal has been examined for pin-hole leaks, the capsule is dried and weighed again.

† Luth, Jahns, and Tuttle (1964) and Burnham and Jahns (1962) describe the technique for this type of experiment. Robertson and Wyllie (1971a, b) have evolved a classification for experiments ranging from 'dry' to water-saturated.

‡ A small funnel may be used to avoid excessive spillage of powder.

Externally heated pressure-vessels

If no significant weight loss has occurred, the capsule is placed in a drying oven at 110 °C for a few hours before final weighing. This indicates if water has been lost during welding.

Before placing it in the pressure-vessel, the capsule is squeezed flat to prevent it jamming on the walls of the vessel should it burst during the experiment, and to allow more than one capsule to be placed in the vessel. The pressure vessel is cleaned with a wire brush and the sample placed in the bottom. If the temperature of the experiment is less than 700–800 °C, up to three capsules may be placed alongside one another in a standard $\frac{1}{4}$-in i.d. vessel; at higher temperatures, the capsules tend to stick together and only one should be put in each vessel. The vessel is filled with distilled water and the back-up rod inserted (Fig. 5.2(b)). A small amount of MoS_2 lubricant is smeared on the threads of the vessel. The cone seat of the closure is properly seated and the closure nut screwed on by hand before tightening it with a wrench, using the minimum necessary force. The vessel is attached to the pressure line and the required pressure applied. Normally the pressurized vessel is left for an hour or so to ascertain there are no leaks. During this period, the furnace should be raised to the required temperature, and, if an external recording thermocouple is being used, it should be inserted into the thermocouple well.

The vessel is then placed in the furnace. As the temperature rises, the excess pressure should be bled from the vessel by 'cracking' a dump valve in the system. During run-up and quenching procedures, no sudden changes in pressure should take place, or the pressure differential between capsule and vessel may rupture the capsule, ruining the experiment. When the vessel has reached pressure–temperature equilibrium, final temperature and pressure adjustments are made before the valve isolating the vessel from the rest of the system is closed (Fig. 5.5).

During the experiment, temperature and pressure should be checked regularly, either by automatic recording devices or manually. If each vessel in the system has its own pressure gauge (Fig. 5.5), the valve isolating the vessel will normally remain closed throughout the run. If, however, the pressures in several vessels are being recorded on a single gauge, the pressure must be checked by equalizing the pressure in the line and vessel and 'cracking' the valve isolating the vessel from the line. If pressure is being maintained in the vessel, no fluctuation in the pressure recorded by the gauge should be apparent.

In most hydrothermal experiments, quenching is done at run pressure. This procedure can be dangerous and the precautions noted later in this chapter should be observed. The first step is to equalize pressure in the pressure line and vessel before opening the valve connecting the two parts of the system (Fig. 5.5). The vessel is then removed from the furnace and cooled initially by applying a jet of air to the bottom of the vessel for a few minutes, followed by a water quench for the final cooling. Quenching may be done by remote control, which is much less dangerous. The rate of air cooling is recorded, and, when the vessel reaches about 200 °C, the water quench may be applied by placing the vessel in a bucket of water, or by using a stream of water from a polyethylene squeeze bottle. If the experiment does not require very rapid quenching, the vessel can be quenched in air alone. During quenching, pressure is maintained by mechanical or manual pumping. This is simplified if a high-pressure reservoir is incorporated in the system. When the vessel reaches room temperature, the pressure is released by opening the dump valve and the vessel is unscrewed from the pressure line. The closure is loosened and the cone-in-cone seal broken. Care must be taken to break this seal before removing the closure nut, since some residual pressure may remain in the vessel owing to mechanical blockage in the nipple. Finally the filler rod and capsule are removed, and the capsule is dried before weighing. If no leakage has taken place, the weight of the capsule should be unchanged from the pre-run weight. Any capsule which has changed weight should be discarded, since the initial bulk composition may have been lost owing to preferential leaching or addition of material.

The methods described here are only generalized procedures for the operation of cold-seal vessels. Many variations may be necessary according on the conditions of the experiment, the type of vessel used, and, in some cases, the personal preference of the operator.

Pressure-vessel performance

A large number of factors effect the performance of a pressure-vessel. The most important are: the composition and physical state of the alloy; the wall thickness of the vessel; the pressure, temperature, and time conditions under which it is run; the method of quenching; the amount of corrosion from the pressure medium. Normally, failure of pressure-vessels is due to rupture caused by 'creep' rather than to the elastic strength of the metal being exceeded. Fortunately, failure is usually preceeded by swelling of the hot end of the vessel. When this is observed, the vessel should be discarded, or used only cautiously at lower $P-T$ conditions. It is sometimes useful to keep a record of the change in diameter of each vessel. Usually minute cracks start inside the vessel and proceed outwards, eventually reaching the outer wall. Although vessel life varies greatly with running conditions, it is possible to outline the generally expected working conditions of each vessel from the elastic strengths of the alloys used, and, in the case of the commoner

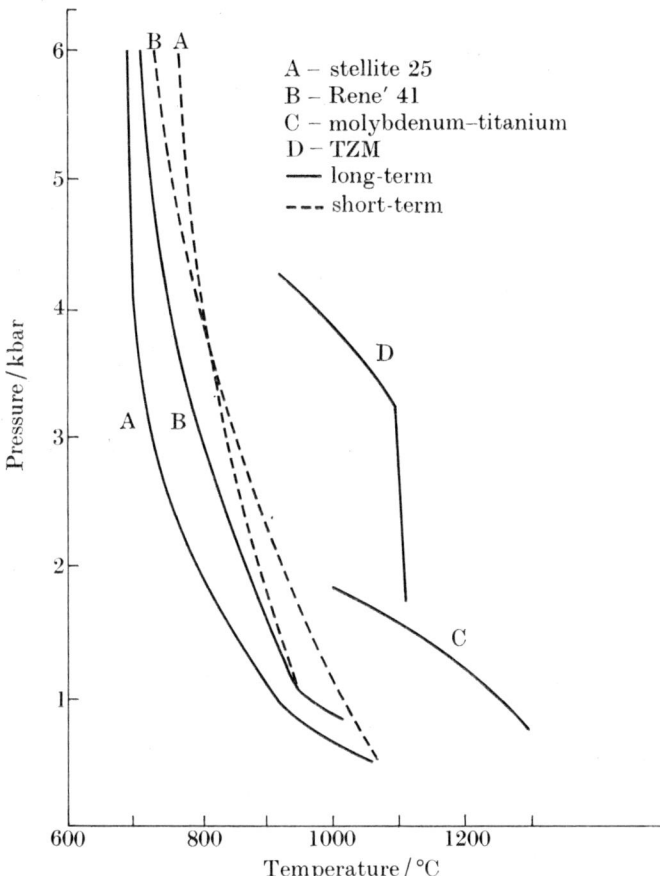

FIG. 5.6. Pressure–temperature limits for various commonly used types of pressure-vessels.

alloys, from experience. Many new alloys have been developed which have expanded the pressure–temperature conditions under which externally heated pressure vessels can be operated.

The maximum operating conditions for some of the commoner alloys are shown in Fig. 5.6. These are based on the stress-rupture curves for long runs (100 hours or longer), and short runs of a few hours, using new, properly hardened and machined vessels with a 4 : 1 wall ratio. As the vessel ages and fractures inside the bore begin to develop, the operating conditions will be lower than those given in Fig. 5.6. This applies particularly to vessels that are repeatedly subject to thermal shock by water quenching.

Pressure-vessels are usually made of one of the following alloys: stainless steel, Stellite 25,† nickel alloys, molybdenum–titanium alloys, and tungsten alloys.

Stainless-steel vessels can be used only for experiments at less than 600 °C at a maximum pressure of 2 kbar. They are, however, inexpensive and are much more easily machined than nickel-based alloys.

Stellite 25 is a cobalt–nickel–chromium-based alloy widely used for pressure-vessels. Vessels of this material with a 4 : 1 wall ratio can be used up to 940 °C at 1 kbar $p(H_2O)$ for short runs, and up to 880 °C at the same pressure for runs of long duration (curve A, Fig. 5.6). At higher pressures, the long-term curve flattens, and above about 3·5 kbar $p(H_2O)$, vessels of Stellite 25 may be used up to 700 °C. The ultimate tensile strength of Stellite 25 rods is about 1100 °C at 1 kbar $p(H_2O)$ to about 825 °C at 4 kbar $p(H_2O)$. Although runs under these conditions are unadvisable, they have been reported (Williams and Harris 1968). The wall ratio in Stellite 25 and other vessels is important, since the larger the ratio the greater the strength imparted to the highly stressed inner wall by the less highly stressed outer wall. Wall ratios of less than 4 : 1 are not normally used.

Nickel-based alloys, such as Rene 41‡ and Nimonic 105,† can be operated at maximum temperature 50–60 °C higher than Stellite 25, as shown in Fig. 5.6 (curve B) for long-term runs. For short-term runs above about 3·5 kbar $p(H_2O)$, the curves for Stellite 25 and Rene 41 cross (Fig. 5.6), indicating that Stellite 25 is superior under these conditions. Although not included in Fig. 5.6, Nimonic 105 has properties similar to those of Rene 41.

Luth and Tuttle (1963) have used Rene 41 with a 5 : 1 wall ratio for vessels capable of operating at 10 kbar $p(H_2O)$ at 750 °C for long periods. For this type of condition, the vessels must be reheat-treated regularly. As described above, Fawcett et al. (1971) have improved Luth and Tuttle's design. Althaus (1969) has used a hemispherical closure for Rene 41 and Nimonic 105 vessels with a 7 : 1 wall ratio for long experiments at 750 °C and 12 kbar $p(H_2O)$. Above 7 kbar, argon must be

---

† Registered trade marks of Haynes Stellite Co.
‡ Registered trade mark of Henry Wiggin and Co.

used as an external pressure medium, since water freezes at room temperature under this pressure. Althaus (1969) avoids freezing at high pressures by adding 50 per cent by volume of ethylene glycol to the water.

Recently, higher-strength nickel-based alloys have been developed which may be suitable for even higher temperatures. Two of these, Nimonic 115 and UCAR M22† seem particularly promising.

Vessels made of molybdenum alloys (Fig. 5.6, curves C and D) considerably extend the temperature limits of externally heated pressure

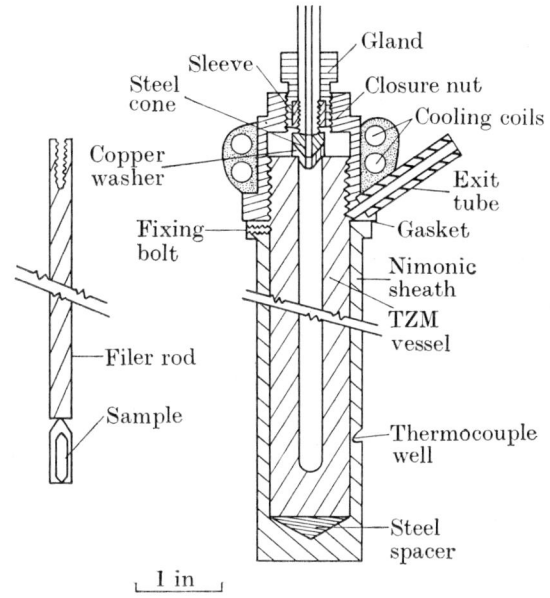

FIG. 5.7. T–Z–M-type pressure-vessel. (After Williams 1968.)

vessels up to about 4 kbar $p(H_2O)$. Unfortunately, these vessels are more difficult to operate than those described previously since they tend to oxidize in air and water at high temperatures, and because of differences in thermal expansion between the vessel and the stainless-steel closure.

Williams (1966, 1968) and Williams and Harris (1968) have described molybdenum alloy vessels and have tested their performance between 900 °C and 1300 °C at various pressures. Williams (1966, 1968) has designed a pressure-vessel (Mo–0·5%Ti) with a 3:1 wall ratio for operation at 1200 °C at 1 kbar $p(H_2O)$ (Fig. 5.7). To avoid oxidation, argon is

† Registered trade mark of International Nickel Co.

used as an external pressure medium and a sheath of Nimonic 75 alloy protects the vessels from air. In his earlier design (Williams 1966), the sheath was not firmly attached to the vessel, and the air between the vessel and sheath was expurged by passing argon, at a pressure slightly higher than atmospheric pressure, into the space between the vessel and sheath. In his 1968 design, Williams eliminates this space by placing a tightly fitting sheath around the vessel, attached by two bolts at the top of the vessel. A soft gasket cement provides an airtight seal with closure nut (Fig. 5.7). Alternatively, the sheath may be silver-soldered and bolted to the closure nut. The spacer at the bottom of the sheath minimizes the volume of trapped air, which escapes through the exit tube on heating. This tube is closed when the temperature of the experiment is reached.

Williams (1968) also describes a vessel made of strain-hardened TZM alloy (molybdenum–0·5% titanium–0·08% zirconium) with a 4:1 wall ratio which can be operated at high temperatures and much higher pressures than his previous design (Fig. 5.6, curve D).

For molybdenum alloy vessels, it is necessary to keep the stainless-steel closure nut at constant temperature, because the much smaller coefficient of thermal expansion of molybdenum relative to stainless steel will otherwise result in leakage during cooling. This problem can be eliminated by placing water-cooling coils around the closure as shown in Fig. 5.7.

A further difficulty with molybdenum vessels is the tendency of the molybdenum to alloy with the sample container. This can be avoided by placing the sample capsule in a loosely fitting nickel or platinum sheath attached to the bottom of the filler rod. Insertion and removal of samples is facilitated by a gland nut and nipple which screw into the top of the closure nut. This arrangement permits insertion and extraction of the sample without removal of the closure nut, which is bolted to the sheath (Fig. 5.7). A threaded filler rod allows easy extraction. A copper washer prevents the steel closure cone being scored by the much harder molybdenum alloy.

Although not yet extensively tested, arc-melted tungsten alloys may be used for externally heated pressure-vessels to extend the temperature to about 1500 °C at 2 kbar pressure (Williams and Harris 1968). Despite the high cost of such alloys, they are probably cheaper and less troublesome than the internally heated vessels currently required for these conditions.

Special uses of externally heated pressure-vessels

In addition to the 'standard' techniques for $p(H_2O)-T$ determinations described above, externally heated pressure-vessels are used for many specialized experiments. Some of these, such as experiments for controlling partial volatile pressures, require only modification of procedure; others, such as the Shaw-type vessels for controlling oxidation by introduction of hydrogen through a permeable membrane (Shaw 1967), also require modification of the vessel. Both of these are described in Chapter 8.

Externally heated pressure-vessels are used extensively for solubility experiments and for solubility–precipitation experiments. For this type of work a large volume of solvent must be analysed, and, consequently, large-capacity Morey-vessels are often used or Tuttle-type vessels of up to $\frac{1}{2}$ in or 1 in i.d. and correspondingly larger o.d.

Barnes (1963) has designed a vessel, termed a 'rocking autoclave' (or Barnes 'bomb'), for carefully controlled measurement of solubility of gases or liquids. In this apparatus, weighed amounts of solid, liquid, or vapour may be introduced into the vessel, and vapour and coexisting liquid may be extracted from the vessel for analysis, without changing the $P-T$ conditions of the vessel. Modifications of large-volume Morey vessels have also been adapted for solubility measurements (Currie 1968). Solubility work with corrosive liquids can be done in Tuttle-type vessels fitted with gold, silver, or platinum liners.

Extended length pressure-vessels are also used for experiments requiring thermal gradients such as solubility–precipitation work or metasomatism experiments. In the latter case double-closure vessels may be used. These permit flow of the pressure medium through the systems as well as a temperature gradient. Barnes (1971) describes a small-volume (50-ml) system for equilibrating solids up to 600 °C at 1 kbar $p(H_2O)$.

Large vessels are often used for preparation of batches of starting-material, such as glasses, which must be used in a crystalline or semi-crystalline state for quenching experiments. Some glasses (see Chapter 3) cannot be readily crystallized 'dry' and must be crystallized hydrothermally in a sealed capsule.

Externally heated vessels are also used for $P-V-T$ measurements (Presnall 1969, 1971).

Hazards and precautions

Operation of externally heated pressure-vessels involves many potential hazards, some of which have already been mentioned. Among the dangers are failure of vessels during quenching, caused by thermal shock; failure due to improper machining or heat-treatment of vessels, closures, etc.; failure caused by overheating of furnaces; poorly designed running conditions; and explosive escape of the hot pressure medium from pressure tubing, valves, etc.

The most dangerous stage of hydrothermal experiments is during quenching. When the hot, pressurized vessel is out of the furnace, the operator may be exposed to the major hazard of an exploding vessel or to the minor hazards of a failure in pressure tubing, which can result in his being sprayed by a fine powerful jet of hot water. It is good practice to quench vessels by remote-control methods in which the operator is protected by a shield, and furnaces, etc., are raised and lowered by a pulley arrangement. Such an arrangement is mandatory for high-pressure equipment.

As mentioned above (p. 110), nickel-based alloys tend to swell before rupture by creep. Williams and Harris (1968) have investigated pressure-vessel failure and conclude that large wall ratios and increasing ductility of the alloy produce greater swelling before failure. The most dangerous materials for pressure-vessels are molybdenum and tungsten, which are brittle and can cause explosive release of pressure. Such an occurrence exposes the operator to fragments of the vessel and filler rod moving at the speed of bullets.

The rate of cooling, or the frequency of rapid quenching (i.e. directly into water) will also increase the rate of failure of vessels. This should be avoided wherever possible.

Removal of the vessel from the pressure line and removal of the closure nut must also be done cautiously. The pressure in the vessel must obviously be released before removing the vessel from the line. Both the connecting and closure nut should be loosened slowly in case of inherent pressure build-up due to blockage. During all quenching procedures safety glasses should be worn and exposed parts of the body protected. Even under 'normal' conditions, pressure-vessels tend to 'spit' hot water.

Rupture of a vessel during 'run-up' or during the experiment is less likely to cause damage, since the vessel is usually protected by fairly thick furnace insulation which absorbs flying fragments. However, danger still exists from filler rods, particularly if the vessel is operated in a horizontal position. Although the loading and unloading of such vessels may be awkward, horizontally operated vessels should have their closures toward a wall and should be protected by shielding at all times.

Such shielding can be cheaply constructed from $\frac{1}{4}$–$\frac{1}{2}$-in boiler plate lined with $\frac{1}{4}$-in plywood to absorb flying particles and prevent ricocheting.

Proper machining and heat-treatment of vessels are also essential. The depth of the axial hole and the thermocouple well should be exact, since any errors may cause them to 'join up'. Eccentricity of the axial hole or sharp corners at the base of the cavity will result in premature failure. It is also essential that the material of the closure nut be properly machined and hardened; the nut may otherwise crack even at low temperatures, resulting in the explosive ejection of the filler rod. If corrosive solutions are used they may readily attack nickel alloys (Morey and Hesselgesser 1952) and the stainless-steel parts of the closure nut. The latter can be protected by frequent resurfacing and coating the nut with MoS_2 grease before each run. The lives of Rene 41 or Nimonic pressure-vessels may be increased by reheat-treating (Luth and Tuttle 1963).

One of the commonest causes of pressure-vessel rupture is the overheating of furnaces owing to failure of the temperature controller. All controllers used in pressure experiments should be fitted with thermocouple break protection which automatically turns the power 'off' at a temperature only a few degree above the designated temperature. In this regard, one of the most frequent causes of failure of temperature controllers to regulate properly is the oxidation of control thermocouples. These should be frequently inspected, both for deterioration and for their correct positioning with respect to the hot-spot in the furnace.

High-pressure tubing and valves may also be potentially dangerous. Bending of rigid pressure tubing is to be avoided (see p. 103); excessive bending may cause minute ruptures and eventual failure. Capillary pressure tubing is particularly dangerous since it may 'whip' if a leak develops. A convenient protection against this is to enclose capillary pressure tubing in spirally-wound flexible metal tubes. Dangerous leakage at valves and connectors is less likely but it may occur and adequate protection should be provided.

Probably one of the greatest hazards in any pressure laboratory is complacency, leading to carelessness. The experimenter should be aware that the slang term 'bomb' for 'pressure-vessel' is used with considerable justification. A few burst 'bombs' lying around any laboratory of this type will reduce complacency to a minimum.

References

ALTHAUS, E. (1969). *Neues Jahrb. Mineral. Abhard.* **111**, 74.
BARNES, H. L. (1963). *Econ. Geol.* **58**, 1054.

BARNES, H. L. In *Research techniques for high pressure and high temperature* (ed. G. C. Ulmer) Springer–Verlag, New York, 317.
BURNHAM, C. W. and JAHNS, R. L. (1962). *Am. J. Sci.* **260**, 721.
CURRIE, K. L. (1968). *Am. J. Sci.* **266**, 321.
FAWCETT, J. J., DAVIES, I. and JAMES, R. S. (1972). *Miner. Mag.* **38**, 529.
GORANSON, R. W. (1931). *Am. J. Sci.* **22**, 481.
GORDON, T. M. and GREENWOOD, H. J. (1971). *Am. Miner.* **56**, 1674.
GREENWOOD, H. J. (1962). *Carnegie Inst. Wash. Year Book* 61, 82.
—— (1967a). In *Researches in geochemistry II* (ed. P. H. Abelson), J. Wiley, New York, 542.
—— (1967b). *Am. Miner.* **52**, 1669.
HADIDIACOS, C. G. (1969). *J. Geol.* **77**, 365.
HOLLOWAY, J. R., BURNHAM, C. W. and MULLHOLLEN, G. L. (1968). *J. Geophys. Res.* **73**, 6598.
KENNEDY, G. C., 1950. *Am. J. Sci.* **248**, 540.
LUTH, W. C., JAHNS, R. H. and TUTTLE, O. F. (1964). *J. Geophys. Res.* **69**, 759.
—— and TUTTLE, O. F. (1963). *Am. Miner.* **48**, 1401.
MOREY, G. W. (1918). *J. Engineers. Club, Philadelphia*, 1.
—— (1953). *J. Am. Ceram. Soc.* **36**, 279.
—— and FENNER, C. N. (1917). *J. Am. Chem. Soc.* **39**, 1173.
—— and HESSELGESSER, J. M. (1951). *Econ. Geol.* **46**, 821.
—— —— (1952). *Am. J. Sci.*, Bowen vol., 343.
—— and INGERSON, E. (1937). *Am. Miner.* **22**, 1121.
PRESNALL, D. C. (1969). *J. Geophys. Res.* **74**, 6026.
—— (1971). In *Research techniques for high pressure and high temperature* (ed. G. C. Ulmer), Springer–Verlag, New York, 259.
ROBERTSON, J. K. and WYLLIE, P. J. (1971a). *J. Geol.* **79**, 549.
—— —— (1971b). *Am. J. Sci.* **271**, 252.
ROOYMANS, C. J. M. (1967). Ph.D. thesis, University of Amsterdam.
SHAW, H. R., 1967. In *Researches in geochemistry II* (ed. P. H. Abelson), p. 521.
TUTTLE, O. F. (1948). *Am. J. Sci.* **246**, 628.
—— (1949). *Bull. geol. Soc. Am.* **60**, 1727.
WILLIAMS, D. W. (1966). *Miner. Mag.* **35**, 1003.
—— (1968). *Am. Miner.* **53**, 1765.
—— and HARRIS, P. G. (1968). *Proc. Instit. Mech. Engineers*, **182**, pt. 3C, 166.
WYLLIE, P. J. (1966). *J. geol. Education*, **14**, 93.

6. Internally heated pressure-vessels

Introduction

As the name implies, internally heated pressure-vessels have the furnace within the pressurized vessel, the wall of which is water-cooled to maintain the strength of the metal. Such vessels can be used at higher pressures and temperatures than externally heated types since they do not depend on the hot-rupture strength of the metal used in their construction but are limited only by the furnace windings, the insulation, and the wall thickness. Internally heated vessels are much larger both in overall dimensions and pressurized volumes than externally heated ones because of the necessity to incorporate the furnace, sample containers, etc. This larger volume makes them particularly useful for investigating reactions with a fluid phase. Internally heated vessels may be operated in the range 5–30 kbar, but for most geological purposes they are rarely used below 5 kbar or above about 10 kbar. Externally heated vessels are normally used up to 5 kbar, and solid-media apparatus, described in Chapter 7, is used for pressures from 10 kbar to about 60 kbar. The equipment described in this chapter is suitable for experiments at pressures from 5 to 10 kbar at temperatures up to 1550 °C.

Because of their larger gas volumes and the necessity of taking several electrical leads through the heads of internally heated pressure-vessels, they are more complicated, dangerous, and expensive to operate than externally heated vessels. Accessories, particularly high-pressure intensifiers and valves, contribute to this expense and their operation requires an efficient operator and a well-equipped machine shop.

Detailed descriptions of internally heated pressure-vessels and their accessories are unfortunately scarce. Smyth and Adams (1923) described an internally heated vessel, which was later modified by Goranson (1931) for his pioneering experiments on silicate solubility. This vessel was cumbersome and difficult to use. Yoder (1950a) described and designed a more practical vessel which has been used for many experiments at gas pressures up to 10 kbar. In the 1960s several other systems were described (Goldsmith and Heard 1961; Burnham 1962; Burnham, Holloway, and Davis 1969). Holloway (1971) has given an excellent description of the design and operation of internally heated vessels, based largely on Burnham's design. This chapter describes Yoder's (1950a), Goldsmith and Heard's (1961), and Burnham *et al.* (1969) vessels, and one developed by the Harwood Engineering Co. Inc. of Walpole, Mass.

Internally heated pressure-vessel designs

General principles

Internally heated pressure-vessels for use in the 5–10 kbar range are constructed from a single block of vacuum-melted tool steel, heat-treated to exactly the correct hardness. A sketch of a typical vessel is shown in Fig. 6.1. Some commonly used steels are Vascojet (1000),†

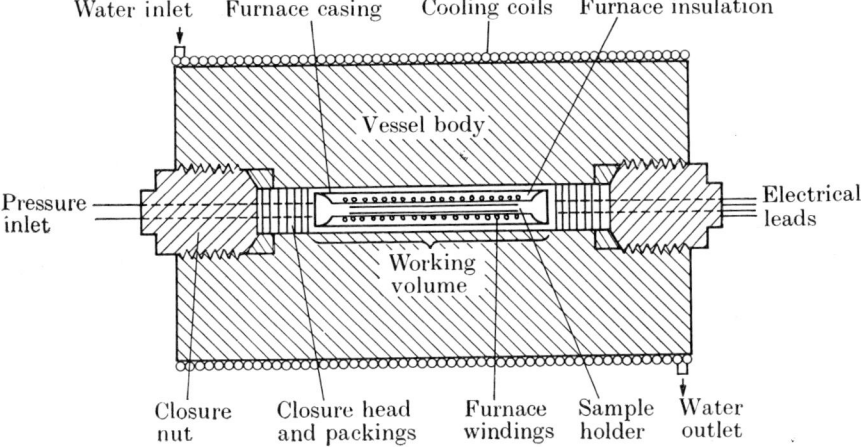

FIG. 6.1. Sketch of simplified internally heated pressure-vessel.

Vascomax (300),† or tool steel SAE 4340. The recommended hardness for the latter is Rockwell C–40 to 42 (Holloway 1971). It is essential that the steel block is of uniform hardness and ductility, and is free from defects.

The dimensions of these vessels are variable but the greater the inside diameter the larger the outside diameter, and hence the cost of the vessel. The maximum pressure attainable depends on the type of steel used and the ratio of outside to inside diameters. Generally a 6 : 1 diameter ratio is allowed. With an inside diameter of 2·5–5 cm an outside diameter of 15–30 cm is thus required. The length of a vessel is normally about twice its diameter. The length of the vessel depends on the desired working volume within the cavity. Most of this working

† Trademark Vasco-A Teledyne Co., Latrobe, Pennsylvania.

volume is taken up by the furnace and insulation, the length of which depends on the maximum operating temperature and length of the hot spot required. Water is circulated through copper coils wrapped around the outside of the vessel; this maintains the strength of the interior of the vessel when operated at high temperatures.

Vessels for the 5–10-kbar pressure range are normally open at both ends (Fig. 6.1) but can be operated with only one open end. The chief disadvantage of the single open-ended vessel is that both electrical and pressure leads must pass through the one closure head. Each end of the vessel is closed by a closure nut, closure head, and closure packings. The closure nut is made of heat-treated stainless steel and supports the closure head and packing. Usually the threaded closure nut has a diameter slightly less than one-half the outside diameter of the vessel and threads to a depth of 15–20 cm depending on the length of the vessel. Thus in the initial forging of an open-ended vessel, the threaded cavities at either end are made considerably larger than the pressurized (working) cavity. The closures contain coaxial holes through which electrical and pressure leads pass into the working cavity. Electrical leads, transmitting power to the furnace, require large holes since they must carry high currents; thermocouple lead holes are much smaller. Bleed holes in the closure nut facilitate detection of pressure leaks.

The remainder of the pressure vessel consists of the pressurized working cavity containing an independently controlled two- or three-zone furnace and its insulation. The insulation is extremely important since it determines the maximum temperature for a given resistance wire, the size and mobility of the hot-spot, and the thermal gradient in the furnace. At high pressures, the thermal gradient can be very large owing to convection currents in the hot gas. All 'unused' space in the cavity should be occupied in order to reduce the volume of the gas. The furnace and thermocouple leads pass through the closure nut and are connected to the external power supply and temperature recorder and controller.

A sample container, made of a suitable material (molybdenum or platinum is often used), separates the sample capsules from the furnace windings. Two or more recording thermocouples are placed close to the sample capsules to monitor temperature gradients. Another thermocouple is placed close to the furnace windings to control temperature. The sample capsules are often noble metal tubes prepared as described in Chapter 5.

Details of the method of pressure sealing, the techniques for passing electrical leads into the vessel, the furnace design and insulation, and the thermocouple systems are discussed in detail in the following sections.

Two basic methods are used to maintain pressure in internally heated vessels; in both gas pressure is generated by a hydraulically operated intensifier or intensifiers (see p. 131). In the first method the vessel is connected directly to the final-stage intensifier, and therefore both vessel and intensifier must be kept at the same pressure throughout the run, although the high-pressure parts can be isolated from the low-pressure parts of the pumping system. Holloway (1971) refers to this method as the integral intensifier system. The advantage of this technique is that no high-pressure valve between intensifier and vessel is necessary. Two disadvantages are that high pressure must be raised with a single stroke of the intensifier, and that each vessel must have its own pumping system. This method is more suitable for vessels of small volume such as those of Yoder (1950a), Birch, Robertson, and Clark (1957), and Newhall and Abbot (1968).

In the second method, a high-pressure valve separates the last stage of the intensifier from the vessel. This method has been referred to by Holloway (1971) as the separate intensifier system. It has the advantages of allowing multiple stroking of the intensifier to reach the pressure required and of allowing several vessels to be operated from a single pumping system. Its major disadvantage is that the high-pressure valve separating the vessel from the pumping system may fail. This method is more desirable for vessels of larger volume and is used in the designs of Burnham et al. (1969) and in the vessel made by the Harwood Engineering Co. Inc.

Yoder's (1950a) internally heated pressure-vessel

A typical example of a small volume internally heated vessel, incorporating an integral intensifier system, was described by Yoder (1950a). A schematic diagram of this vessel and pumping system is shown in Fig. 6.2. Maximum operating conditions are 1400 °C at 10 kbar.

Pressure is attained by a two-stage hydraulically operated gas intensifier, using argon as the pressure medium. Argon is commonly used in internally heated pressure-vessels because it does not react with the metal of the vessel, diffuses through the vessel very slowly, has a low compressibility, and can be obtained in a high degree of purity. At pressures greater than 12 kbar, argon solidifies at room temperature and all pressure lines have to be fitted with heating coils. Nitrogen is occasionally used but has less desirable properties.

The lower part of Fig. 6.2 represents the gas portion of the pressure system; the upper part, the hydraulic portion. By opening the appropriate valves, argon at approximately 150 bars is allowed to pass into the first- and second-stage compression cylinders (A and C respectively in Fig. 6.2) and into the vessel (B). Needle-type valves, rated at 4 kbar,

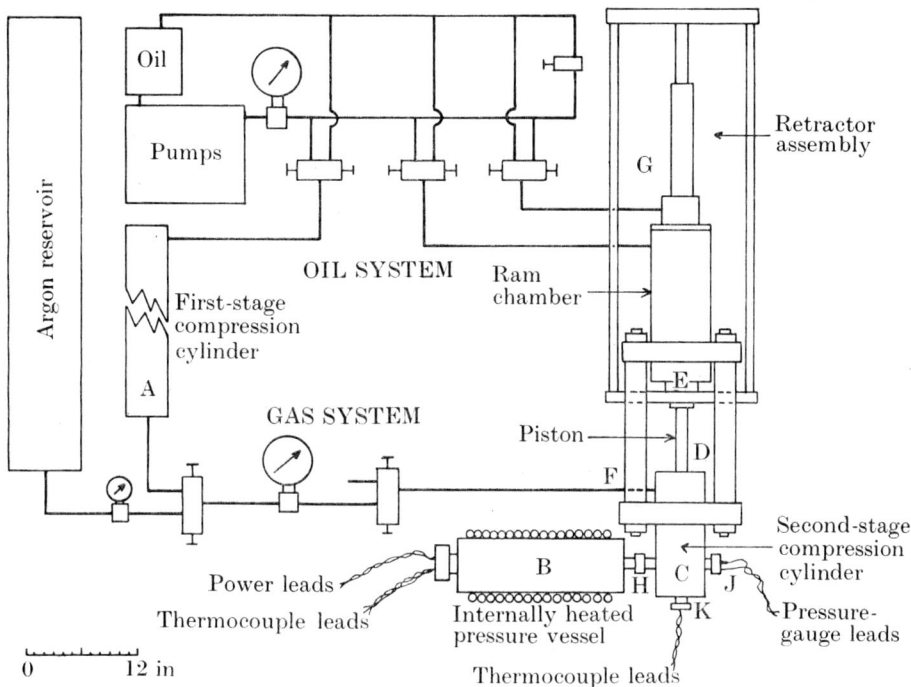

Fig. 6.2. Internally heated pressure-vessel and pumping system. (After Yoder 1950a.)

separate the argon tank and the first- and second-stage compression cylinders (A and C). Oil is pumped into the top of cylinder A, forcing a piston down and pressurizing the gas in B and C to 2 kbar. Packings on the piston separate the oil from the gas. Initially compression rates are slow and may be increased by using a large-capacity pump up to 400 bars. Low-pressure measurements are made with gauges of Bourdon-tube type (Fig. 6.2).

To raise the pressure in the vessel B from 2 kbar to about 10 kbar, oil is pumped into the ram chamber, forcing ram (E) and connecting piston D into the second-stage compression system C. The ratio of the cross-sectional areas of ram E and piston D is 15:1, producing a theoretical pressure of 30 kbar. In practice, this cannot be attained owing to limitations in space and strength of the vessel and connectors, and because of friction between the ram and piston. As the piston D passes the top of the compression cylinder C it closes the port F, thus sealing parts A to F of the system from the high-pressure (i.e. greater than 2 kbar) portion. The piston D can be raised at the end of a run, or in the event of a leak, by activating the opposing cylinder G and releasing the gas opening the valve to the right of the pressure gauge, as shown in Fig. 6.2.

The vessel B is 18 in long with an o.d. of 6·5 in. The working cavity is 0·75 in in diameter and 8 in long. Copper cooling coils surround the outside of the vessel, one end of which is connected to the second-stage

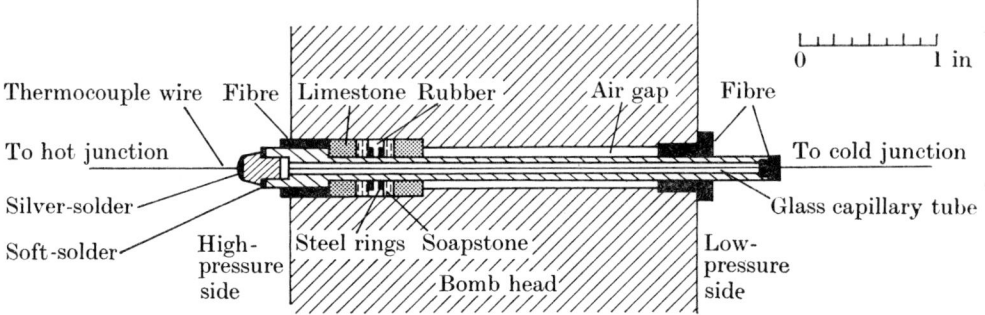

FIG. 6.3. Method of sealing leads through closure heads. (After Yoder 1950a.)

compression cylinder C by a connector H, through which the pressurized gas enters the vessel. The power and thermocouple leads are brought out of the opposite end of the vessel through gas-tight seals.

For most internally heated pressure-vessels, pressure is maintained by a Bridgman unsupported area-type seal. The principles of these seals, and the various modifications are based on original designs described by Bridgman (1949), Bradley and Munro (1965), Weale (1967), and Holloway (1971) These seals are discussed in a later section of this chapter.

In Yoder's design, two electrically insulated, gas-tight thermocouple leads and one power lead pass through one of the closures of the vessel. The vessel itself is used as the earth (ground) lead for the furnace. The method of ensuring electrical insulation and gas pressure sealing is shown in Fig. 6.3, and is based on a design of Birch (1932) with packings by Adams, Williamson, and Johnston (1919). For both thermocouple and power leads, a heat-treated (Rockwell C–40) SAE 4140 steel rod, 3·75 in long, is used. One end of the rod is machined very slightly smaller than the diameter of the hole through the closure head; the other end is about two-thirds of this diameter. For the thermocouple leads (see Fig. 6.3) each rod is carefully drilled, from both ends, to give an i.d. of 0·0625 in. Details of this technique are given in Yoder (1950a). A capillary glass tube is inserted in this diameter to insulate rod and thermocouple wire. The large end of the rod is tinned and a series of packings, consisting of limestone, soapstone, steel, rubber, steel, soapstone, and limestone placed on the rod (Fig. 6.3). The rubber O-ring is oversized and must be punched into position when the rod is inserted in the closure head. The assembly is then arbor-pressed into the seat. Fibre insulation supports the rod at both the high- and low-pressure ends of the closure, with an air gap between the limestone packing and the fibre at the low-pressure side of the closure.

The thermocouple wire is inserted through a 0·0136-in hole drilled in a small mushroom-shaped button which is soldered on to the high-pressure side of the closure head. The wire is silver-soldered to the button and passed through the rod. Details of this rather difficult operation are given by Yoder (1950a). Under pressure, the solder behaves as a soft gasket. When the closure nut is tightened, the rings and packings provide a gas-tight seal with electrical insulation being provided by the capillary glass tubing. Two minor, but detectable, sources of e.m.f. occur across the button owing to surficial alloying in the wire where it is soldered, and to stress gradients in the vicinity of the button. Yoder (1950a) reports 1 megohm resistance between the vessel head and each lead. In order to maintain high resistance under humid conditions, the electrical head should be desiccated when not in use.

Thermocouple and power leads passing through the closure are connected to the furnace within the vessel, and to the power supply and temperature controller outside the vessel. A single-zone platinum furnace wound on a thin-walled ceramic tube, 6 in in length and of $\frac{1}{2}$ in i.d. is suggested by Yoder (1950a). This is cemented into a stainless-steel tube with alumina cement.† Fired soapstone is used to eliminate all free space in and around the furnace and thus minimize temperature gradients due to convection of the hot gas. Porous insulating materials are of little use for this purpose since the gas rapidly penetrates all dead space within the pores.

With one control and one recording thermocouple, Yoder determined a hot spot of ± 5 °C over 2 in long at 800 °C. This temperature required approximately 1 kW of power. Schairer (1959) states that Yoder also

† The cement recommended is RA–98 (Norton International).

126 Internally heated pressure-vessels

used a tungsten-wound furnace for temperatures of 1550 °C at 10 kbar. No details of these windings are given.

For his experiments on the high–low quartz inversion, Yoder (1950a) used a nickel block approximately $\frac{1}{2} \times \frac{5}{8}$ in as sample holder. Within this block, temperatures varied less than 0·1 °C. A platinum sample holder can also be used in this design.

Pressure is measured to within ± 10–20 bars at 10 kbar by a silk-covered manganin cell placed in the second-stage compression cylinder (Fig. 6.2, J). The cell consists of 0·004-in diameter coiled wire with approximately 100 Ω resistance. The coils were calibrated using the method of Bridgman (1911, 1914). The lead of the pressure cell is brought out of the compression cylinder through an insulated plug (Fig. 6.2). The other lead is silver-soldered to earth. Pressure is measured by determining the change in resistance of the manganin coil using a modified Wheatstone bridge as described by Adams et al. (1919). Since the pressure coefficient of manganin wire may change with temperature, a thermocouple is passed through the plug K (Fig. 6.2) to measure any increase in coil temperature during an experiment. The slight temperature increase on the coil has negligible affect on the pressure.

Goldsmith and Heard's internally heated pressure-vessel

Goldsmith and Heard (1961) describe an internally heated pressure-vessel (based on an earlier design of Professors D. T. Griggs and H. C. Heard) with an integral intensifier pressure system for operation at 10 kbar and 1200 °C. Two notable features of this design are its small size (pressure-vessel and accessories can be placed in an area 2·5 by 4 ft), and the rapid method of sample exchange, pumping, and heating.

Goldsmith and Heard's vessel is shown in Fig. 6.4. It is open at both ends as well as along its length. The vessel, forged from Bethlehem Omega tool steel hardened to Rockwell C-52, is of 5 in o.d., and 17 in long, with a working cavity 1 in in diameter and 7 in long. Copper cooling coils are placed on the outside of the vessel. Heating is by a single-zone $Pt_{90}Rh_{10}$ furnace, wound on a fused silica tube 6 in long and 9 mm o.d., with a wall thickness of 1 mm. The windings, 0·025 in in diameter, are cemented on to the silica tube, fired, and machined to 0·96 in diameter. A stainless steel jacket of 0·015 in wall thickness is placed over the furnace and secured to the main unsupported area packing closure (Fig. 6.4).

One of the power leads is grounded to the steel jacket, the other attached to an 8° tapered tool-steel plug. This plug is insulated with a 0·010-in wall pyrophyllite core, providing an electrically insulated, leak-tight pressure seal through the centre of the main vessel closure (Fig. 6.4). The low-pressure end of the plug is connected to the power supply through a large-capacity variac and constant-voltage transformer to minimize voltage fluctuations.

Four thermocouple leads, one common and three live, pass through holes in the main closure, equidistant from the central power lead. The

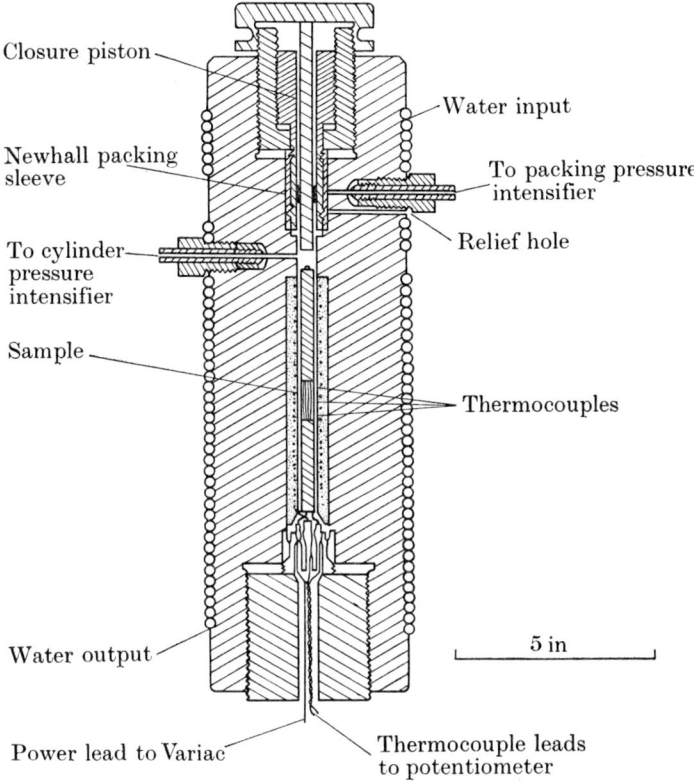

FIG. 6.4. Internally heated pressure-vessel. (After Goldsmith and Heard 1961.)

thermocouple wires are threaded through steel cones, with axial holes 0·0135 in in diameter, and silver-soldered on the high-pressure side in a manner similar to the power lead. The wires for each of the three thermocouples are sheathed in double-bore fused silica tubing covered with platinum-foil, and placed ⅜ in apart midway along the centre of the furnace. The sample is contained in a holder of slightly smaller diameter than the furnace bore. This holder telescopes over the two thermocouples showing lowest thermal gradient and is readily removed through the

Newhall packing closure, thus eliminating the necessity of removal of the main closure except for maintenance purposes (Fig. 6.4). Temperature gradients between adjacent thermocouples are normally less than 5 °C (Goldsmith and Heard 1961, p. 47).

All free space within the pressurized working cavity of the vessel is filled with 1-mm o.d. fused silica rods. Goldsmith and Heard (1961) report that at 1200 °C and 10 kbar the fused-silica furnace core tends to devitrify, producing spalling and eventual failure of the furnace.

The main difference between Goldsmith and Heards' internally heated pressure-vessel and Yoder's design is in the adaptation of the Newhall controlled clearance packing around a closure piston for one of the closure heads (Fig. 6.4). The principles of the controlled clearance packing are given by Johnson and Newhall (1953). It eliminates the necessity for unsupported area type packings in the closure head. A rapid method of pressure generation is, however, required for maximum efficiency.

Sealing of the upper closure (Fig. 6.4) is made solely by the radial elastic distortion of a packing sleeve around the well-fitting closure piston. Gas-tight packing is placed between the piston and sleeve at either end of the sleeve, and a suitable gas, such as petroleum ether, under high pressure produces elastic distortion of the packing sleeve. This gas is pumped into the vessel through an annular hole in the vessel wall. The packing pressure gas is separated from the main cylinder fluid by another gas-tight packing, and a relief hole drilled through the vessel wall prevents excessive pressure build-up of the gas, which might cause intermixing of the two pressurizing systems (Fig. 6.4). A 0·2-in wide step, half-way up the sleeve, provides contact between the closure piston and sleeve. The contact between the sleeve and closure packing is critical. Goldsmith and Heard (1961, p. 48) recommend an 8–12 microinch ground-surface finish on both piston and sleeve and a clearance of 0·0002–0·0005 in between them. Using these values, the ratio of packing to cylinder pressures of 0·8 will maintain cylinder pressures between 3 and 10 kbar of argon. Provided that the contact surfaces are kept clean and the packing sleeve is not permanently distorted by excessive packing pressure, the piston-sleeve closure can be used for many experiments.

The hydraulic pressure-generating system used by Goldsmith and Heard is shown in Fig. 6.5; it supplies both the packing and cylinder pressures. As in Yoder's system, a two-stage gas-compression technique is used. Argon, at tank pressure, passes through the system to the vessel. Initial compression is achieved by a separator which ensures isolation of the kerosene oil pumping medium from the argon. This is done by making the piston in the floating separator double-ended and fitting both ends with Bridgman mushroom-type packings (Bridgman 1949). The unsupported area is opened to the atmosphere by drilling holes through the piston and separator wall, resulting in a pressure of 2 kbar being rapidly attained which passes through valve H (Fig. 6.5) into the confining pressure and packing, pressure intensifiers, and vessel. The

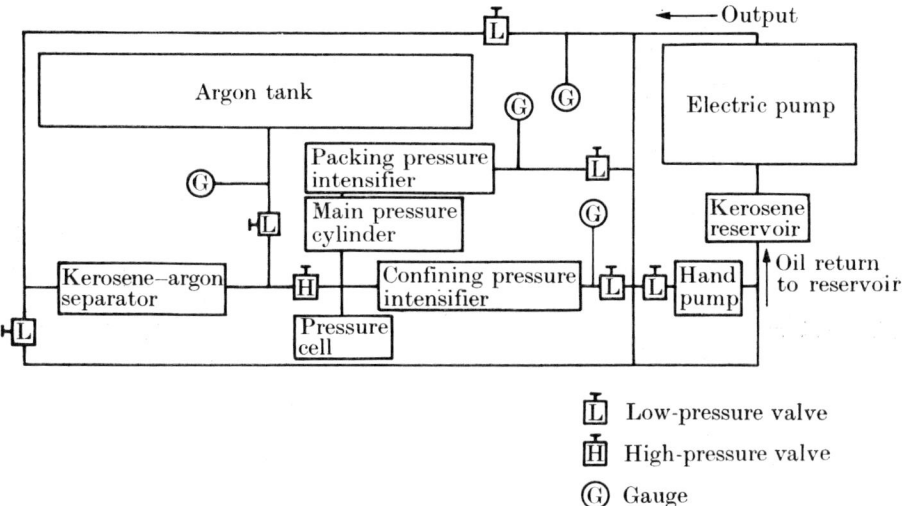

FIG. 6.5. Pumping system for vessel shown in Fig. 6.4. (After Goldsmith and Heard 1961.)

kerosene and oil are returned to the reservoir. The second compression stage for the fluid in the confining-pressure intensifier is achieved with a 15 : 1 ratio intensifier driven by a commercial 50-ton ram, producing pressures of up to 10 kbar in the cylinder. The gas in the packing pressure intensifier is raised to a lower pressure by a similar 20-ton ram. Details of valves, pressure tubing, etc., are given by Goldsmith and Heard (1961). Confining pressures are measured by a calibrated gold–chrome resistance element immersed in toluene and are estimated to within ± 0.5 per cent.

Harwood Engineering Co. Inc. internally heated pressure-vessel

As an example of a larger-volume internally heated pressure-vessel with a separate pumping system for the 5- to 10-kbar range, the design by Harwood Engineering Co. Inc., Walpole, Massachusetts may be

130   *Internally heated pressure-vessels*

considered. A modification of this design is shown in Fig. 6.6. The vessel is mounted on a trunnion to facilitate operation in a vertical or horizontal position; the vertical position is best for minimizing thermal gradients due to convecting gases. The vessel is open at both ends, closure being made with Bridgman unsupported area type seals. Pressure is applied through one closure and power and thermocouple leads

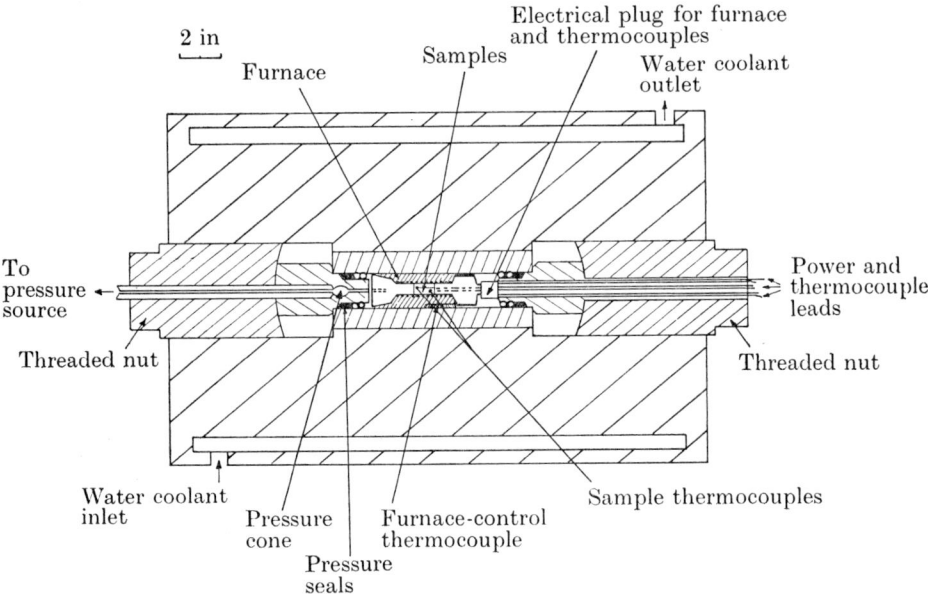

FIG. 6.6. Internally heated pressure-vessel after design of Harwood Engineering Co. Inc., Walpole, Mass.

pass through the other. A total of nine electrical leads may be passed through one closure, permitting insertion of up to four thermocouples and, if desired, power leads for a three-zone furnace. Two or three thermocouples placed at intervals along the sample holder; monitor sample temperatures; the other thermocouple is placed close to the furnace windings and acts as a control thermocouple. In a two- or three-zone furnace, each zone is independently controlled. Both thermocouples and power leads each have a single common lead; thus in a system with four thermocouples only five leads (one common and four live) pass through the head. The general features of multi-zone furnaces, thermocouples, and temperature control, as used in the Harwood design, are considered in a later section.

The method of passing electrical leads through the closure on the Harwood vessel is shown in Fig. 6.7. The insulated power or thermocouple wire passes through a coned beryllium–copper anode. A ceramic insulating back-up bushing is placed on the opposite end of the cone from the anode and a fired crushable-lava cone attached to it. A thin film of a suitable silastic† is smeared between the anode and lava cone. The assemblage is compressed into the closure head with a special pressing tool, and a retaining screw holds the assembly in place. This

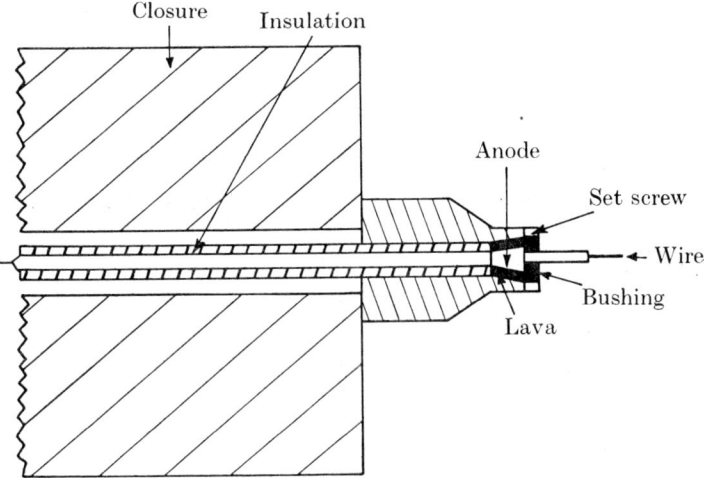

FIG. 6.7. Electrical closure used in vessel shown in Fig. 6.6.

arrangement provides a gas-tight, electrically insulated seal for pressures up to 10 kbar at temperatures in excess of 1200 °C.

The power and thermocouple leads are attached to a core and ceramic socket. The furnace, sample holder, and insulation, all enclosed in a steel casing, are simply plugged into this socket. A locating pin on the socket assures the proper connections between the various electrical leads. With this arrangement, the closure head containing the electrical leads need be removed from the vessel only for routine maintenance.

A schematic diagram of a typical pumping system utilizing a two-stage hydraulically operated gas intensifier is shown in Fig. 6.8. The basic difference between this system and those shown in Figs. 6.2 and 6.5 is that the vessel may be isolated from the pressure generating system by a high-pressure valve (valve D in Fig. 6.8). Consequently several vessels may be operated from one pressure system.

† Silicone Seal made by General Electric Co. is a suitable silastic.

The lower section of Fig. 6.8 represents the hydraulic part of the system; the upper section, the gas part of the system. Argon gas at about 150 bars passes through valve A into the first-stage intensifier, while valve B is closed. The gas is compressed to about 3 kbar in this intensifier and passes into the second-stage intensifier and vessel through valves B, C, and D. During the first-stage compression, which may require several strokes of the pump, valve A is closed. With valve D closed, the first-stage intensifier is isolated from the rest of the system and the gas

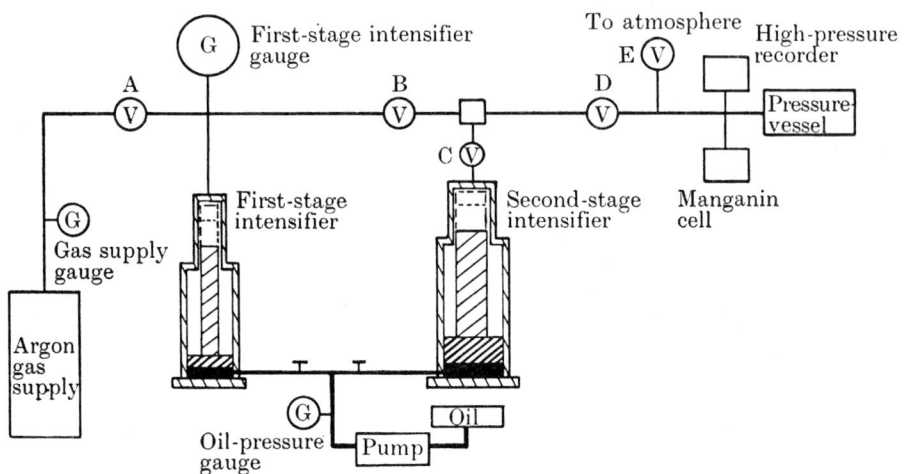

FIG. 6.8. Possible pumping system for vessels shown in Figs. 6.6 and 6.9.

in the second stage is compressed from about 3 kbar to a maximum pressure of about 13 kbar and passes into the vessel through valves C and D. When thermal equilibrium is attained, valve D is closed and the pressurized vessel is isolated from the pumping system. With valve C closed, the piston in the second-stage intensifier may be retracted to its low-pressure position.

Pressure from the first-stage intensifier is measured with a calibrated Heise gauge. Pressure within the vessel and during the second-stage compression is measured by a manganin cell and continuously recorded. The manganin cell may be placed outside the vessel (Fig. 6.8) or, if required, inside the vessel. In the latter case a further lead must pass through the closure of the vessel.

Details of the intensifier packings, high-pressure tubing, and high-pressure valves are given in a later part of this chapter.

Internally heated pressure-vessel of Burnham, Holloway, and Davis

In 1969, Burnham *et al.* described an internally heated pressure-vessel for operation to 10 kbar at 1500 °C. This design, which is an adaptation of one described by Burnham (1962), consists of two interconnected vessels and has been used especially for $P-V-T$ gas measurements

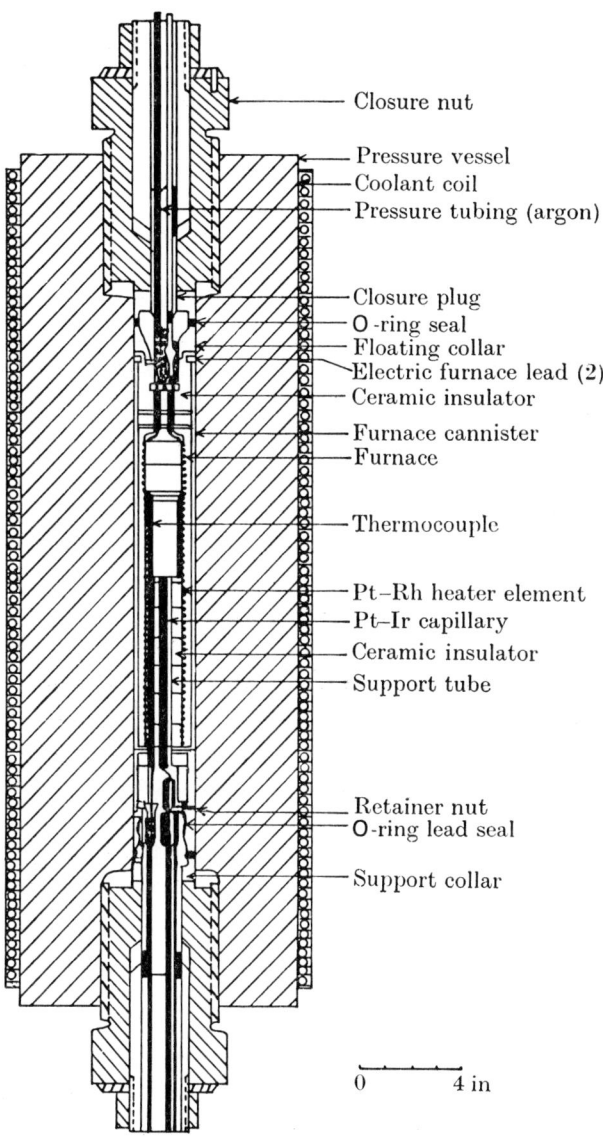

FIG. 6.9. Internally heated pressure-vessel. (After Burnham *et al.* 1969.)

(Burnham *et al.* 1969) and $P-V-T$ measurements on hydrous silicate melts (Burnham and Davis, 1971). This vessel has also been used as a model by Holloway (1971) in his description of internally heated pressure-vessels.

The vessel, shown in Fig. 6.9, is open at both ends and operates on the separate intensifier pumping system. Vacuum-melted steel (modified SAE-4340), heat-treated to Rockwell C-40-42, is used for each vessel. The dimensions are 12 in o.d., 2 in i.d., and 28 in in length; the working cavity is approximately 16 in long and accommodating a furnace and a $P-V-T$ cell. The second pressure-vessel used by Burnham *et al.* (1969), for investigation of the specific volume of water at high pressures, contains a linear variable differential transformer and a bellows. Details of the second pressure-vessel and auxiliary equipment are given in Burnham *et al.* (1969).

The closure plugs are sealed by double O-rings of different sizes which fit against different-sized diameters of a hardened steel collar (Fig. 6.9). Since this collar is free to move, a differential force is exerted on the larger rubber O-ring proportional to the pressure within the vessel. This differential force produces radial pressure of the O-ring against the vessel wall and is independent of hydrostatic pressure. The O-rings are held in place by mild steel anti-extrusion rings. The innermost O-ring is supported by a lead washer which maintains the seal at high pressures when the O-ring shrinks.

For their experiments on $P-V-T$ relations of water, Burnham *et al.* (1969) used three electrical leads and one pressure lead on each closure. No details are given by these authors of the method of passing these leads through the closure head, but Holloway (1971, pp. 233–6) describes various sealing methods suitable for this type of vessel. With some modification, these techniques are similar to those described previously for the Harwood vessel.

For pressure generation, Burnham *et al.* use a two-stage gas-intensifier system, as described above. Pressures are measured by a manganin cell and continuously recorded.

Bridgman unsupported area seals

Various forms of the Bridgman unsupported area seal are used for internally heated pressure-vessels and accessories. The use of this type of seal has been discussed above in describing pressure-vessels; a detailed review of the principles and operation of the unsupported area seal, including various modifications, is given by Holloway (1971). The term 'unsupported area' is derived from the fact that the stem of the closure plug is not dependent on the closure nut for support. The theory is discussed by Holloway (1971).

For large-diameter internally heated vessels, a modification termed the *inverted Bridgman seal* is most commonly used. This seal makes removal of the closure plug much simpler than with the regular unsupported area seal, which tends to deform the pressure-vessel wall.

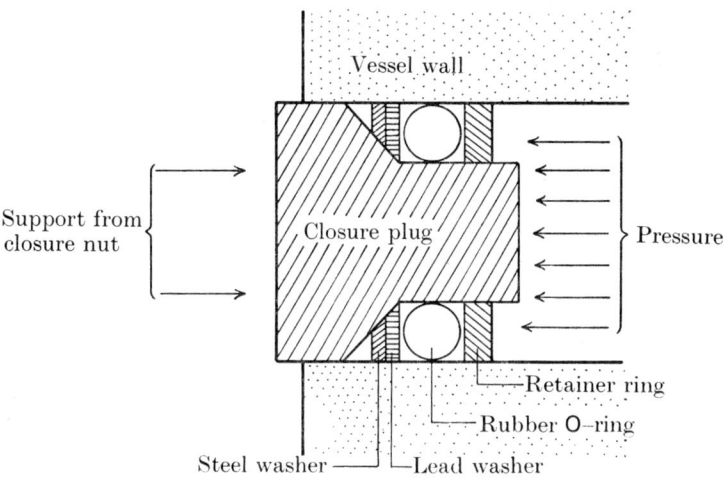

FIG. 6.10. Inverted Bridgman unsupported area seal. (After Holloway 1971.)

A sketch of the inverted Bridgman seal is shown in Fig. 6.10. The closure plug is supported by a closure nut and a series of rings. The inner bevelled washer of hardened steel fits the bevel of the closure head. Outwards from this washer, a lead ring, a rubber O-ring, and a retainer ring complete the sealing arrangement. The diameter of the O-ring is slightly larger than the inside diameter of the vessel and provides the initial seal by squeezing it between the closure plug and vessel wall. As gas pressure increases, the O-ring is forced against the lead ring (Fig. 6.10), which deforms and is squeezed against both vessel wall and closure plug. At this stage the increased gas pressure produces shrinkage in the O-ring, allowing gas to leak past the ring against the lead washer which is now deformed. The force on the lead washer is transmitted to the inner steel washer. An unsupported area exists where the steel ring is not in contact with the bevelled part of the closure plug. The pressure within this unsupported area is greater than the gas pressure on the steel ring, which is forced against the wall of the vessel.

A version of the unsupported area seal developed by Professor C. W. Burnham is described by Holloway (1971).

Accessory equipment

Of utmost importance in the successful operation of internally heated pressure-vessels is the maintenance and design of accessory equipment. In this section some of the basic principles of pressure intensifiers, valves, high-pressure connectors and tubings, manganin cells; as well as furnace insulation and windings, temperature control, and measurement are considered. Maintenance of this high-pressure–high-temperature equipment requires a skilled and resourceful machinist, since many of the parts cannot be bought from stock and must be made. An excellent review of accessory equipment is given by Holloway (1971).

Pressure equipment

Pressure intensifiers. Because of the need to apply pressure to relatively large volumes which are heated to high temperatures, gas is the only suitable pressure medium for internally heated vessels. The choice of gas is limited by its compressibility, chemical inertness, etc. Most operators use argon, although, as noted above (p. 100) it solidifies at room temperature at pressures above 12 kbar.

Pressure intensifiers compress a large volume of gas at low pressure to a small volume of gas at high pressure. This is done by a large hydraulically driven piston, operating at low pressure, which forces a smaller piston to pressurize the gas. The difference in the ratio of the areas of the small and large piston is termed the *intensification ratio* and determines the maximum gas pressure. If a separate intensifier is used for pressures in the 5–10 kbar range, two or more intensifiers may be used in tandem. The first-stage intensifier is designed for large volumes of gas producing a low pressure up to about 3 kbar; the second, smaller-volume, intensifier produces higher pressures owing to the decreasing compressibility of the gas with increasing pressure. With this type of arrangement, the initial compression can be done with two to three strokes of the first-stage intensifier, and the pressure raised to about 10 kbar with the second-stage intensifier. Total pumping time is about 15 to 20 minutes.

The design and operation of intensifiers is reviewed by Newhall (1957) and by Manning and Labrow (1971). A schematic diagram of a two-stage gas intensifier (from the design of the Harwood Engineering Co. Inc.) with details of the packings is shown in Fig. 6.11. The basic difference between the low-pressure (first-stage) intensifier and high-pressure (second-stage) intensifier is that in the former the piston is both raised and lowered by the hydraulic oil, whereas in the latter the hydraulic system is raised by the oil but lowered by the gas pressure. The intensification ratio in the first stage is 28·5:1, producing a maximum pressure of 50 000 lb in^{-2}; the ratio in the second stage is 130:1, producing a maximum pressure of 200 000 lb in^{-2} argon gas.

The first-stage intensifier contains three sets of packings. The lowest set, consisting of low-pressure Teflon V-ring packings (Fig. 6.11), separates the hydraulic oil from the ram assembly. The middle set of packings prevents the gas from leaking between the sides of the ram and cylinder wall into the lower-pressure hydraulic oil (Fig. 6.11, inset). They consist, from bottom to top, of a brass wedge ring, a Teflon packing, a rubber packing washer, and brass wedge rings. At the upper part of the ram packings, a brass wedge ring, a Teflon back-up ring, a rubber O-ring, and a brass packing are used. The uppermost set provide the high-pressure closure at the top of the cylinder. These packings are placed between the tapered closure nut and intensifier wall; they consist, from bottom to top, of a steel retaining ring, a brass packing ring, an O-ring, a Teflon non-extrusion ring, and a brass wedge ring. When the ram reaches the top of its stroke, the gas under pressure passes through a hole in the closure nut 0·1 in in diameter and through high-pressure tubing to the second-stage intensifier (Fig. 6.8).

The second-stage intensifier also has three sets of packings, but these are of different design because of the difference in size and intensification ratio. The lower set of packings (Fig. 6.11), separating the hydraulic oil from the piston, do not contain V-packings as in the first-stage intensifier, since the oil in this case only drives the piston up-stroke and is not involved in returning the piston to its low-pressure position. In the second-stage, the packings on the hydraulic part of the system are O-rings, a brass collar, and a wedge packing. A further difference between the first- and second-stage intensifiers is in the ram. As shown in Fig. 6.11, the narrow part of the ram is housed in, but not connected to, the lower wider part of the piston. The middle packings (Fig. 6.11, inset), located at the top of the ram, consist of alternating brass and Teflon packings. The top set of packings (Fig. 6.11, inset) consist of a steel retaining ring, a steel closure packing ring, an O-ring, a Teflon non-extrusion ring, a lead ring, and a steel wedge ring placed between the

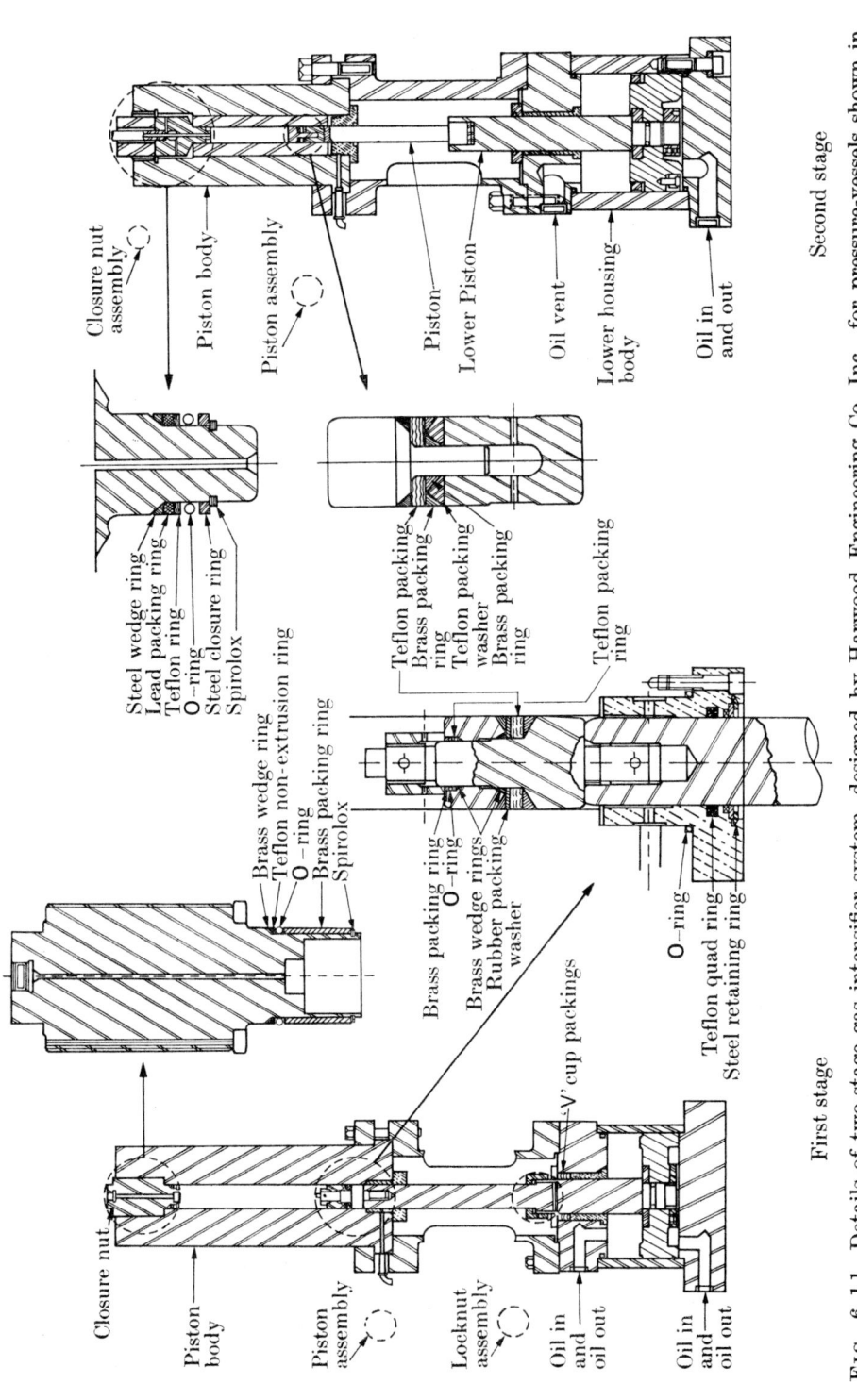

Fig. 6.11. Details of two-stage gas intensifier system, designed by Harwood Engineering Co. Inc., for pressure-vessels shown in Figs. 6.6 and 6.9.

intensifier wall and closure plug. The closure plug is partly supported by the closure nut; the arrangement is similar to that of the modified inverted Bridgman seal discussed by Holloway (1971). The gas, now under high pressure, passes into the vessel.

To regulate the rate of pumping, the system is equipped with a throttle valve. Rupture discs, rated for various pressures, are placed at strategic points in the system to prevent explosion due to valve leakage.

The pressure system described here operates extremely well with a minimum of maintenance. Certain routine precautions to avoid problems should, however, be observed. It is essential that the correct oil be used in the hydraulic system and that all foreign particles larger than 10 μm be removed. Cleanliness is most important in all high-pressure systems; small particles will score pistons, packings, etc. For this reason, the hydraulic system should be primed and purged of air at low pressure before operation. Both metallic and non-metallic packings require periodic renewal. All worn packings of organic material, or metal packings that are scratched, must be replaced. Care must be taken when the cylinder heads are removed to avoid dust falling into the cylinder, and the piston and intensifier walls should be thoroughly cleaned before repacing.

Pressure tubing and fittings. For pressures below about 5 kbar, 'standard' types of pressure tubings and fittings as described in Chapter 5 are used. For pressures in excess of 5 kbar and up to about 13 kbar, 316 stainless-steel tubing $\frac{3}{16}$ in o.d. and 0·025 in i.d. is used. Alternatively, a composite tubing with a 316 stainless-steel core and an alloy-steel casing may be used. Tubing of this type, $\frac{3}{4}$ in o.d. and $\frac{1}{16}$ in i.d., has a higher rupture strength than the single 316 stainless-steel tubing for the same working pressure. The number of fittings, valves, etc., should be kept to a minimum, since they provide potential points of leakage. Fittings are connected to pressure tubing by gland nut and tubing collar assemblies (see Holloway 1971).

High-pressure valves. For pressures less than about 4 kbar, the valves described in Chapter 5 may be used (Fig. 6.8). For higher pressures up to 13 kbar, valves which are hand-tightened are no longer suitable since the correct torque required for sealing cannot be gauged by 'feel'. Normally such valves are seated by a torque wrench or a spring, set to supply the correct seating force.

A schematic diagram of a needle valve with a torque-limiting handle for service to about 13 kbar is shown in Fig. 6.12.† Sealing is accomplished

† From the design of the Harwood Engineering Co. Inc.

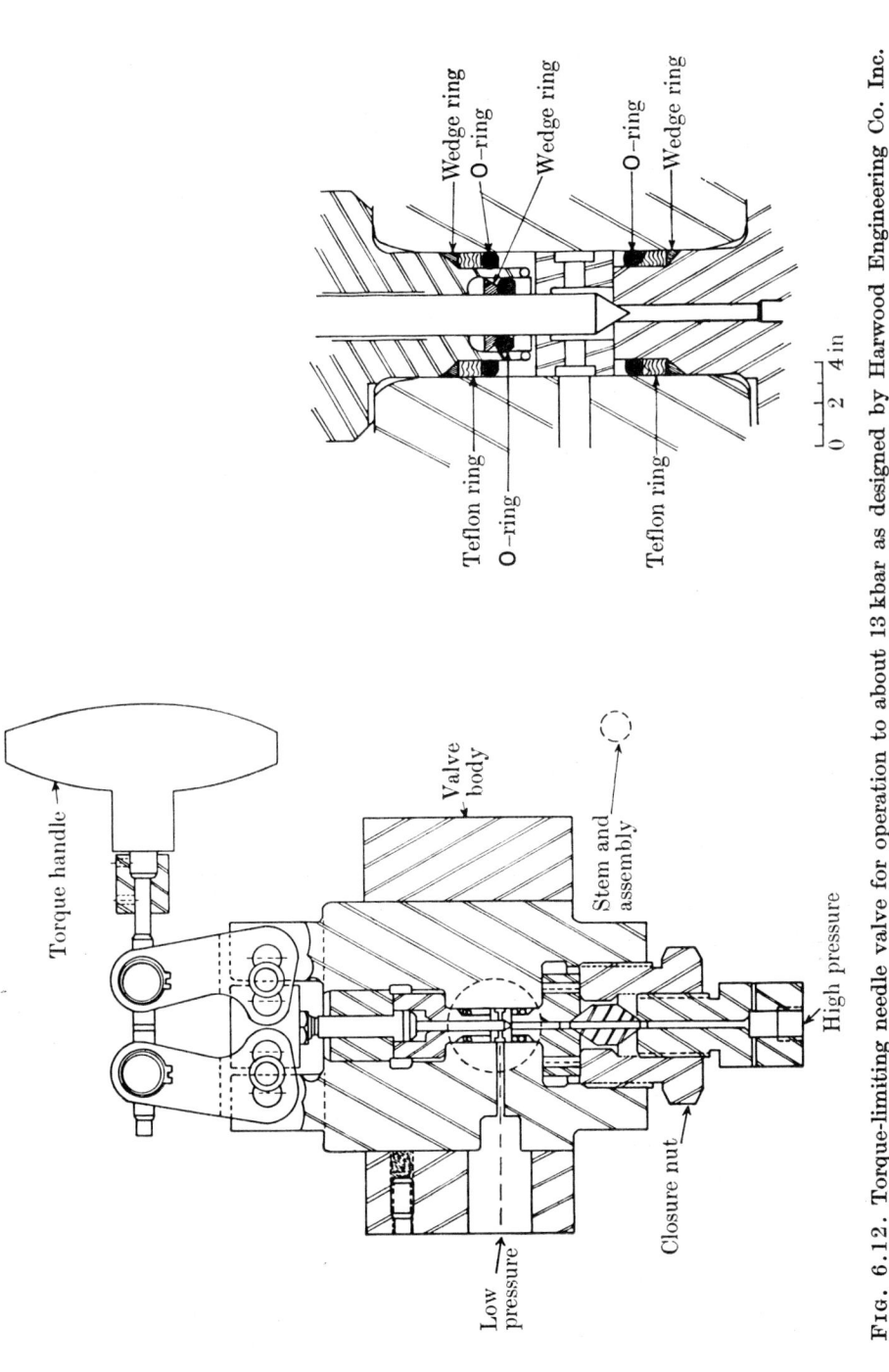

Fig. 6.12. Torque-limiting needle valve for operation to about 13 kbar as designed by Harwood Engineering Co. Inc. Details of needle valve shown on right-hand side.

by an unsupported area type packing, details of which are shown in Fig. 6.12. The high-pressure seal is maintained by the tip of the needle valve seating into the closure head of the valve body. A wedge ring, a Teflon ring, and an O-ring are placed between the closure heads and valve body (Fig. 6.12). On the low-pressure side, the gas is prevented from escaping past the valve stem by an O-ring and wedge-ring seal. To provide the correct load of the valve tip of the seat, and to prevent the valve stem being broken by overtightening, a torque-limiting handle is incorporated in the valve stem. For the design shown in Fig. 6.12 the required torque is 35 ft-lb.

An alternative design and a general discussion of the problems of high-pressure valves are given by Holloway (1971).

Pressure measurement. For pressures less than about 7 kbar, Bourdon-tube gauges are sufficiently accurate and are inexpensive. Above this pressure, manganin cells are commonly used. These consist of two non-inductively wound coils, one consisting of a manganese alloy wire whose resistance is sensitive to pressure change; the other, a temperature-compensating coil. Both coils are attached to a Wheatstone bridge circuit. Manganin coils must be calibrated periodically against a dead-weight piston gauge, or a secondary standard such as a calibrated gauge.

Leak detection. One of the more frustrating problems of high pressure research is the detection of small gas leaks. Closure nuts and other pressure fittings are equipped with 'bleed' holes through which gas, leaking past pressure packings, can be detected. If the leak is large, the escaping gas may be audible or detected by the presence of a bubble in a non-corrosive soap–water mixture. Electronic gas-leak detectors, which measure thermal conductivity, are also available. These instruments can detect argon leaks as low as 4.5×10^{-4} cm^3. Detection of leaks by addition of radioactive ^{85}Kr to the argon gas also provides a sensitive method (Spetzler *et al.* 1969).

Temperature equipment

Furnaces. Attaining high temperatures in internally heated pressure-vessels is more difficult than in equipment in which the furnace is not under pressure. This is due to the presence of gas, which affects the electrical properties of the resistance elements; the fact that smaller furnaces are (generally) required than for typical externally heated pressure-vessels; the much larger volume of gas in internally heated vessels; and the necessity of controlling thermal gradients caused by gas convection. These factors are considered before discussing furnace design.

Furnaces are heated by helically-wound resistance wires, which are basically the same as described in Chapters 4 and 5. For temperatures less than 1200 °C, base metal elements can be used; and for higher temperatures platinum (or platinum alloys) or tungsten windings are required. In the presence of gas under pressure, the electrical properties of these wires alter drastically. For example, the hot resistance (R_h) of 0·02-in diameter Pt wire at 1000 °C in air at atmospheric pressure is roughly one-third of the calculated value at the same temperature and 5 kbar of argon. Hence the power requirements for internally heated pressure-vessel furnaces are much greater than those for externally heated systems. Vodar and Saurel (1963) indicate the power required with a nitrogen pressure medium for various pressures and temperatures. Argon is believed to have similar requirements. Data of Goldsmith and Heard (1961, p. 47) indicate that the amount of power required to maintain 1100 °C at 1 kbar argon pressure (0·9 kW) is proportionately larger than at 10 kbar (1·7 kW).

Pressurized gas also causes fluctuation in the hot-spot of the furnace, owing to changes in the density of the gas with temperature and pressure. Variation of 2–3 kbar may shift the hot-spot by 2–3 cm. This situation may be improved by designing a two- or even three-zone furnace in which each zone is independently controlled. This design also serves to minimize thermal gradients and provides a larger hot-spot. Ideally, a multi-zoned furnace would be desirable but the limitations in the number of power leads that can be taken through the vessel closure make this impracticable.

The maximum temperature produced in any furnace depends on the diameter and length of the windings. In the relatively long furnaces used for externally heated systems the length of the tube is adequate for the desired diameter and length of the windings, but in internally heated systems the length of the furnace tube may be insufficient for the necessary windings. Although the selection of diameter, total length, and type of wire to give optimum temperatures is complex, some general rules should be observed. The larger the wire diameter the less is the likelihood of breakage through oxidation or brittleness. The spacing between adjacent windings should not, however, be less than the diameter of the wire, and hence the larger the wire diameter the shorter the total length of the windings. Some compromise between wire diameter and wire length is often necessary.

If noble metal wires are used, the cold, and hence hot, resistances may vary greatly depending on whether pure metal or noble metal alloys are used; increasing amounts of the second metal do not necessarily produce a uniform change in resistances. For example, pure platinum wire, 0·040 in in diameter has a cold resistance (R_c) of 0·040 Ωft^{-1}, whereas $Pt_{90}Rh_{10}$, $Pt_{80}Rh_{20}$, and $Pt_{60}Rh_{40}$ wire of the same diameter have resistances of 0·072, 0·078, and 0·066 Ωft^{-1} respectively. Whatever wire is used, the manufacturer's specified resistances should be carefully observed to calculate the optimum diameter, wire length, and current density before designing the furnace. Formulae for these calculations are given in the *Handbook of Chemistry and Physics* and in wire manufacturer's catalogues.

The maximum possible amount of insulation, occupying all free space in the working cavity, should be used in internally heated pressure-vessel furnaces in order to minimize thermal gradients and obtain maximum temperatures. Various types of insulation have been used. Yoder (1950b) showed that material with large amounts of porous dead space is rapidly penetrated by hot gas at high pressures and soon becomes ineffective. Pyrophyllite (or fired soapstone) has been used (Yoder 1950a). Holloway (1971) and others use layers of silica cloth soaked with an alumina cement slurry. Goldsmith and Heard (1961) advocate fused silica rods as the insulating medium. In addition to the insulation required to avoid gas convection, electrical insulation between the windings and steel furnace casing is required. Alumina cement is normally used for this purpose.

In the design of furnaces for internally heated pressure-vessels two basic rules must be observed: the higher the maximum working temperature the shorter the length of the hot-spot; the greater the furnace insulation, the higher the temperature of the hot-spot.

A schematic diagram of a typical winding arrangement for a two-zone furnace is shown in Fig. 6.13. Wire of the correct type and diameter to provide the necessary temperature is wound on a coarse-grained highly porous alumina tube in which a groove has been cut lengthwise to accommodate the control thermocouple. The length of this groove should be sufficient to move this thermocouple and locate the hot-spot once this has been determined under various pressure conditions. The high-porosity tube is essential; gas trapped at high pressures and temperatures will be otherwise unable to escape when the experiment is quenched, and the release of pressure will crack the tube.

The wire may be wound on the tube by hand, or mechanically with a lathe set at the desired number of turns per inch. Alternatively, a grooved furnace tube may be used once a satisfactory design has been achieved. After winding, the ends of the wires are tied down. Since the number of power leads which can pass through the closure of the vessel is limited, only one common lead, serving both zones, is used. This is attached at some point along the windings by welding or other means, producing an independently controlled two-zone furnace (Fig. 6.13).

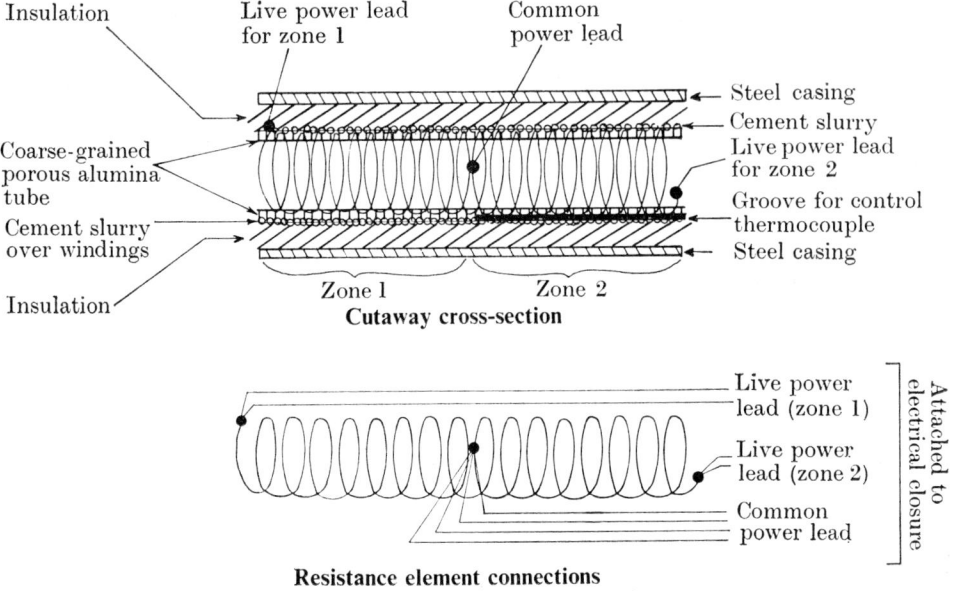

Fig. 6.13. Schematic diagram of winding arrangement for two-zone furnace.

This furnace requires three power leads. For extra strength, the leads are often of double thickness.

A thin coating of alumina cement is applied to the windings and allowed to harden. The tie-wires at both ends of the windings are removed and the wound tube is placed in fired pyrophyllite or other insulating material. The furnace tube and insulation are placed in a thin stainless-steel casing which fits exactly into the working cavity. The power leads are connected to a ceramic connector which fits into a seven-holed plug† at the electrical closure of the vessel. This arrange-

† The number of holes in this plug depends on the number of furnace zones and thermocouples used. A seven-holed plug accommodates a common power and common thermocouple lead, two live power leads, a live control-thermocouple lead, and two live recording-thermocouple leads.

ment permits easy removal and installation of the furnace assembly and is used in the vessel of the Harwood Engineering Co. Inc. described above. Examples of other furnace designs and their construction are given by Holloway (1971).

Thermocouples. At least three thermocouples are desirable in internally heated pressure-vessels: a control thermocouple, placed close to the windings; and two recording thermocouples placed in the sample holder or close to the samples if no holder is used. The relative positions of the thermocouples depend on the design of the furnace and vessel. If the control thermocouple tip is in the centre of the determined hot-spot of the furnace for given pressure–temperature conditions, the recording thermocouples are placed on either side of the control thermocouple. The distance between the tips of the recording thermocouples should be approximately the same as the length of the hot-spot to monitor the hot-spot with changes in pressure, but should not be less than the length of the sample capsule in order that any thermal gradients along the entire sample length may be continuously recorded. The length of the hot-spot varies with furnace design and experimental conditions. Generally a hot-spot of $\frac{1}{2}$ in can be achieved in which variation is not greater than ± 5 °C up to 10 kbar. This allows sample capsules up to $\frac{1}{2}$ in long to be used. An estimation of lateral variation of the hot-spot across the sample holder may be determined by placing one of the recording thermocouple tips close to one of the walls of the cylindrical sample holder, and the other thermocouple tip close to the opposite wall as shown in Fig. 6.14. Lateral temperature variation across the small diameter of the sample holder is generally small.

For temperatures below 1100 °C, Cr–Al thermocouples may be used, but for higher temperatures Pt–Rh couples are required. The Cr–Al thermocouples may be of the sheathed type, using a high-grade ceramic insulator filler. These are superior to the exposed tip type (see Chapter 5) in that the tip is not exposed to the pressure medium. Holloway (1971) recommends an Inconel 600† sheath and describes their construction. Unfortunately sheathed Pt–Rh thermocouples are not practicable owing to the expense and fragility of a platinum alloy sheath.

As for the power leads, a single common lead is used for all thermocouples. The common and live leads are attached to the ceramic connector which fits into the holed plug. From this plug the leads pass through the vessel closure.

Temperature controllers. The recording thermocouple leads are

† Trade mark International Nickel Co.

attached to a continuously recording potentiometer. Power is fed to the furnace through a fully proportionating stepless controller capable of delivering at least 1·8 kW and controlled by the thermocouple close to the furnace windings. In this type of controller, the temperature of each furnace zone is independently controlled by its own variac and the amperage of each winding recorded. A galvanometer circuit records the temperature of the control and hence the temperature of the furnace hot-spot. To prevent overheating owing to failure of the cooling water on the exterior of the vessel, an automatic power shut-off device is incorporated in the temperature controller. This turns off all power to the furnace should the water flow fall below a preset level. A thermocouple-break protection circuit (see Chapter 5) similarly shuts off the power in event of breakage of the control thermocouple.

The effect of argon at pressures up to 7 kbar on the e.m.f.s of Cr–Al and Pt–Rh$_{10}$ thermocouples has recently been discussed by Lazarus, Jeffery, and Weiss (1971). These authors arrive at a temperature correction for Pt–Rh$_{10}$ thermocouples of 0·57–0·03 °C kbar^{-1} in the range 600 °C to 1000 °C. For Cr–Al thermocouples, the pressure correction is very small up to about 800 °C; above this temperature Cr–Al thermocouples rapidly deteriorate, and at 980 °C Lazarus *et al.* suggest that temperature errors with Cr–Al thermocouples in the range of pressure may be as high at 100 °C. The general problems of the pressure effect on thermocouple e.m.f.s are discussed in the following chapter.

Sample holders. The type of sample holder required depends entirely on the nature of the experiment. For example, in his studies of high–low quartz inversion, Yoder used a nickel block containing the quartz sample and an alumina reference (Yoder 1950a, p. 830). Silica tubes may also serve as sample holders. For special applications, no sample holder is necessary provided that the samples do not touch furnace windings or thermocouple tips and remain in the furnace hot-spot.

For experiments with a sealed noble metal capsule containing the reactant and water, a sample holder is normally used. This holder, which usually consists of a cylindrical block of platinum or pure molybdenum metal, contains the sample capsules and recording thermocouples and fits snugly into the furnace tube, occupying all 'free' space within the furnace tube to minimize excess gas convection. Although platinum sample holders are preferable, they are very expensive and molybdenum is perfectly satisfactory provided it is fitted with a platinum liner to prevent oxidation and contamination of the platinum capsules and platinum thermocouple tips.

The sample holder (Fig. 6.14) is slightly shorter than the furnace tube One end is screwed into the pressure end of the furnace. Two holes are bored into the 'electrical' end through which the recording thermocouples pass into the sample cavity. This cavity is normally no larger than $\frac{1}{2}$ in long and $\frac{3}{8}$ in i.d.; the remainder of the holder is a solid block of metal. The holder itself may have a cap at one end with a filler block of metal which screws in between the end of the holder and the sample cavity. Alternatively the holder may be unscrewed along its length, permitting easy insertion and removal of samples. The sample cavity must, of course, coincide with the furnace hot-spot. One of the advantages of

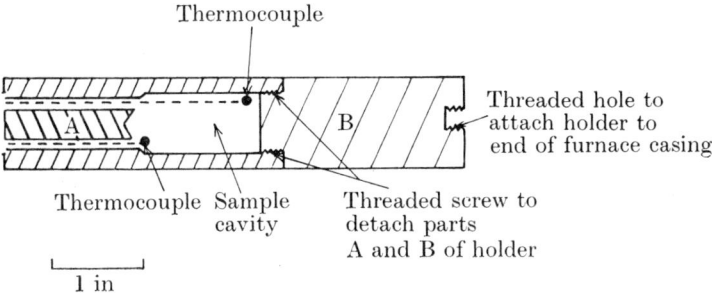

FIG. 6.14. Sample holder for internally heated pressure-vessel.

internally heated pressure-vessels over externally heated vessels is that many more capsules may be accommodated in each experiment. Care must, however, be taken that these capsules do not touch exposed thermocouple tips.

Procedures for internally heated pressure-vessels

The basic principles of operation of internally heated pressure-vessels are similar to those for externally heated types described in Chapter 5. Only the procedures for hydrothermal experiments at high pressures are described here.

Noble metal capsules are filled with the starting-material and distilled water in the required proportions and welded shut in the manner described by Goranson (1931). These are weighed and placed in a pre-dried (110 °C) sample holder and steel-cased furnace. Drying of the furnace is recommended to prevent any moisture being picked up from the air which might cause oxidation of the molybdenum holder or short-circuiting of the furnace windings.

The sample holder is placed in the vessel through the pressure closure, with the vessel in a vertical position, pressure closure end uppermost. This is done by means of an insertion and extraction rod. Care must be taken that the furnace holder plugs into the electrical end with the power and thermocouple leads in their correct holds in the closure plug. This procedure is aided by a locating pin in the closure plug. At this stage, continuity between the electrical end of the vessel and the furnace is checked with a resistance meter.

The pressure closure and head are then inserted, first by hand and then with a large wrench. Excessive pressure must not be used to seat the head, which, when tight, should be 'backed off' about a quarter to half a revolution. Pressure is then applied to a value slightly less than the desired running pressure and the system checked for pressure leaks. If none is detected, water is circulated round the outside of the vessel and the power turned on. For many vessels heating is slow since it takes a considerable time to heat the large volume of gas in the vessel. The pressure and temperature at the recording thermocouples are monitored continuously. Any large variation in the latter indicates a leak in the pressure system or a shift in the predetermined hot-spot of the furnace. In order to keep thermal gradients to a minimum, the vessel should be operated in a vertical position with pressure head uppermost.

When thermal equilibrium has been achieved, any excess pressure in the vessel may be bled slowly. It is advisable to regulate pressure continuously during the run-up procedure to avoid having to pump cold gas into the vessel, which may result in cracking of the furnace tube or other parts. If the pressure must be adjusted during a run, this must be done cautiously to avoid radical changes in the hot-spot position.

At the end of the experiment, quenching is achieved by shutting off the power while maintaining pressure. This results in rapid cooling of the vessel owing to the thermal insulation and the large cold mass of the vessel. Pressure is bled and the closure furnace and sample holder removed. The sample capsules are weighed and examined by optical or X-ray methods, or both. The precautions listed in Chapter 5 are observed to ascertain that the experiment has taken place in a closed system.

Safety precautions

Internally heated pressure-vessel systems are potentially more dangerous than externally heated systems because of their higher pressure and consequently larger potential energy. The most hazardous parts of the system are the smaller parts rather than the vessel itself or pressure intensifiers. Such parts are heavy and, should they fail, their penetrating capacity is small. Power and thermocouple leads, valve stems, and severely bent pressure tubing may fail and be ejected with velocities several times greater than that of a high-powered rifle bullet. The operator must also be constantly aware of the dangers of ricocheting particles.

For the pumping system, boiler plate $\frac{1}{4}$ to $\frac{1}{2}$ in thick lined with plywood should separate the pressurized portions from laboratory staff. Valve stems or high-pressure tubing and fittings should never point directly at the operator. Failure of pressure tubing or a fitting due to a flaw may lead to 'whipping'. This can be prevented by clamping the tubing to rigid material or by using rigid tubular shielding. Operator error within the low-pressure parts of the pressurizing system, may be avoided by incorporating a series of rupture discs set at a safe pressure level. Unfortunately such discs are not entirely reliable at higher pressures. Similar types of shielding must be used for each pressure-vessel, particular precautions being taken at either end of the vessel where leads may blow out. These shields may conveniently be fitted with wheels or suspended from pulleys to facilitate loading and unloading of the non-pressurized vessel.

Overheating through water failure is avoided by incorporating shut-off devices in the temperature controller and thermocouple-break protection devices.

As stressed in Chapter 5, it is most important that staff should be constantly aware of the potential dangers of the type of equipment and act accordingly. Unauthorized personnel should be banned from all high-pressure areas.

References

ADAMS, L. H., WILLIAMSON, E. D. and JOHNSTON, J. (1919). *J. Am. chem. Soc.* **41**, 12.
BIRCH, F. (1932). *Phys. Rev.* **41**, 641.
—— ROBERTSON, E. C. and CLARK, S. P. (1957). *Ind. Engng. Chem.* **49**, 1965.
BRADLEY, R. S. and MUNRO, D. C. (1965). *High pressure chemistry*, Pergamon Press, Oxford.
BRIDGMAN, P. W. (1911). *Proc. Am. Acad. Arts Sci.* **47**, 347.
—— (1914). *Phys. Rev.* **3**, 126.
—— (1949). *The physics of high pressure*, Bell, London.
BURNHAM, C. W. (1962). *Am. ceram. Soc.* 64th annual meeting, Seattle, Washington, Bull. Am. ceram. Soc. **41**, p. xxx.
—— and DAVIS, N. F. (1971). *Am. J. Sci.* **270**, 54.

BURNHAM, C. W., HOLLOWAY, J. R. and DAVIS, N. F. (1969). *Am. J. Sci.* A **267**, 70.
GOLDSMITH, J. R. and HEARD, H. C. (1961). *J. Geol.* **69**, 45.
GORANSON, R. W. (1931). *Am. J. Sci.* **22**, 481.
HOLLOWAY, J. R. (1971). In *Research techniques for high pressures and high temperatures* (ed. G. C. Ulmer), Springer–Verlag, New York, p. 217.
JOHNSON, D. P. and NEWHALL, D. H. (1953). *Trans. Am. Soc. mech. Eng.* **75**, 301.
LAZARUS, D., JEFFERY, R. N. and WEISS, J. D. 1971. *App. Phys. Lett.* **19**, 371.
MANNING, W. R. D. and LABROW, S. (1971). *High pressure engineering*, The Chemical Rubber Co., Cleveland, Ohio, p. 369.
NEWHALL, D. H. (1957). *Ind. eng. Chem.* **49**, 1949.
—— and ABBOT, L. H. (1968). *Proc. Inst. mech. Engrs.* **182**, 288.
SCHAIRER, J. F. (1959). In *Physiochemical measurements at high temperatures*, Butterworths, London, p. 117.
SMYTH, F. H. and ADAMS, L. H. (1923). *J. Am. ceram. Soc.* **45**, 1167.
SPETZLER, H., SCHREIBER, E. and NEWBIGGING, D. (1969). *Rev. sci. Instr.* **40**, 179.
VODAR, B. and SAUREL, J. (1963). In *High pressure physics and chemistry*, vol. 1 (ed. R. S. Bradley), p. xx, Academic Press, London.
WEALE, K. E. (1967). *Chemical reactions at high pressures*, E. and F. N. Spon, London.
YODER, H. S. (1950a). *Trans. Am. geophys. Un.* **31**, 827.
—— (1950b). *J. Geol.* **58**, 221.

7. Solid-media apparatus for pressures above 10 kbar

Introduction

IN recent years there has been a marked increase in experimental results of mineralogical and petrological interest involving both hydrostatic and fluid pressures up to about 60 kbar and temperatures in excess of 1500 °C. Because of its many industrial applications, high-pressure technology is well established and much has been published on this subject. In this chapter only the most commonly used techniques, readily available to earth scientists, which can develop sustained high temperatures and pressures are considered. These include the opposed anvil device (or simple 'squeezer'), the piston–cylinder apparatus, the tetrahedral-anvil system, and the belt apparatus. Of these, the piston–cylinder apparatus, such as is described by Boyd and England (1960a), is most widely used for geological work. The basic designs are modifications, or direct developments of, the work of P. W. Bridgman, summarized in Bridgman (1949) and in his collected papers (1964). More recently several review papers on the applications of high-pressure research in the earth sciences have been published (e.g. Bell 1967; Bradley 1969; Simmons 1968; Wyllie 1963, 1966; Paterson 1970). An excellent historical review of high-pressure research is given by Vodar and Kieffer (1970).

High pressures were originally developed by squeezing the sample between two simple anvils driven by a hydraulic cylinder, thus concentrating the total force over a small area. If the diameters of the anvil and hydraulic cylinder are known, the total pressure amplification can be calculated as the square of the ratios of the diameters. Heat is supplied by an external furnace. If flat anvils and hydraulic cylinders are used, a number of problems arise, not least of which is accurate measurement of pressure and, to a lesser extent, temperature.

The experiments of Bridgman showed that cylinders can withstand much higher compressive stresses if they are tapered and that the wider the angle of taper the greater the compression attainable. As a result, most high-pressure solid-media apparatus is now designed with tapered cylinders, and high temperatures are achieved with internally heated furnaces. A number of modifications of cylinder-type equipment, using several cylinders with different geometric arrangements, have also been developed. In addition, materials capable of very high compressive stresses, such as tungsten carbide, are now available at moderate cost. Indeed, the cost of providing and operating a piston–cylinder device, capable of routine operation up to 50 kbar at 1500 °C, is less than that of an internally heated pressure-vessel for 10 kbar. This is due mainly to the much lower cost of maintaining pressure. Although both types of apparatus require constant attention and the services of a well-trained machinist, routine operation of the higher-pressure equipment is probably easier.

The calibration of pressures and temperatures presents two basic problems in experiments with solid-media apparatus. Pressure cannot be measured by conventional methods but must be computed from the geometry of the apparatus and corrections made for friction and other affects. Although temperatures are measured by thermocouples, they may be in error owing to pressure affects on the e.m.f.s of the thermocouples, as well as to gradients within the sample chamber. Much research has been done within the last decade to determine the magnitude of pressure corrections, using various calibrants and methods of operation of the equipment. The effects of pressure on temperature have also received considerable attention. New thermocouple materials have been investigated and cross-checks between different laboratories initiated.

Examples of three commonly used types of solid-media apparatus for experiments in excess of 10 kbar are here described.

Solid-media equipment

The Bridgman opposed anvil

The opposed-anvil device shown in Fig. 7.1 is based on a design of Griggs and Kennedy (1956) as modified by Dachille and Roy (1960). This is only one of many modifications of Bridgman's original opposed-anvil device, often termed a simple 'squeezer'. Because of the externally heated furnace, the maximum temperature is 700 °C at 50 kbar, although temperatures as high as 1000 °C at lower pressures, or pressures as high as 200 kbar at lower temperatures, are possible (Dachille and Roy 1962).

The essential components of the apparatus (Fig. 7.1) are the hydraulic system, consisting of a 20-ton ram operated by a reciprocating pumping system with a pressure controller and gauge; the pistons, shown in detail in Fig. 7.1(b), one of which is operated by the ram, the other being attached to a platen at the top of a rigid framework containing the entire apparatus. The sample consists of an annular capsule, surrounded by foil a few hundred micrometres in thickness, placed between the two anvils. Dachille and Roy (1963) describe a sealed capsule technique for solid–vapour studies in which the capsule is embedded in pyrophyllite. Thermocouples are placed close to the sample capsule, which is heated by a conventional split-type resistance furnace.

The size of the hydraulic ram, which supplies uniaxial pressure to the pistons and sample, depends on the pressures required and the strength

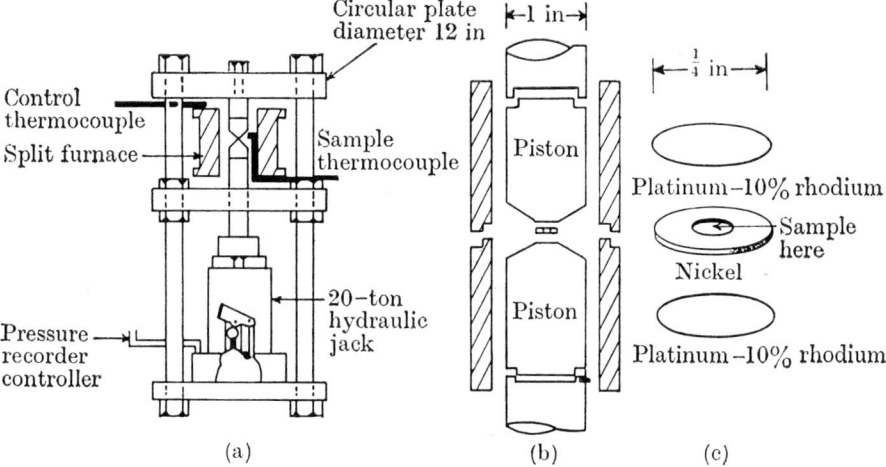

FIG. 7.1 Simple opposed anvil apparatus: (a) entire apparatus; (b) high-pressure pistons; (c) sample assembly. (After Dachille and Roy 1960.)

of the frame. A 20-ton hydraulic ram will produce pressure up to 50 kbar using a 12-in diameter circular platen and a $\frac{1}{4}$-in diameter sample assembly (Fig. 7.1(a), (b), (c)). For work at pressures up to 200 kbar using samples of the same size, a 60-ton ram may be necessary (Dachille and Roy 1963). With such a 'single'-stage device, pressures up to 200 kbar are feasible; and pressures in excess of 400 kbar at low temperatures may be attained with a 'three-stage' device.

The material used for the pistons (anvils) also depends on the desired pressure and temperature conditions. Dachille and Roy (1963, Fig. 2) give the range for various materials. For pressures in the order of 50 kbar, 66 HS or 'Speed Star' alloy is used up to 500 °C, and compound pistons made from tungsten carbide in Rene for temperatures up to about 750 °C at this pressure. For pressures from 60 kbar to about 180 kbar, this material or tungsten carbides in TK tool steel can be used. However, with increasing pressures the permissible temperatures rapidly decrease. Overloading of anvils beyond their recommended limits produces deformation in the steel types and cracking and splintering in those made from tungsten carbide. Steel anvils must be carefully hardened, and the tolerances between anvils with tungsten carbides enclosed within steel are critical. Specifications of these are given by Dachille and Roy (1963).

The length and diameter of the anvils depends partly on the required pressure and sample size. Their length is largely a matter of convenience, and may depend on the size of the hot-spot of the furnace. For the apparatus shown in Fig. 7.1, anvil diameters are 1 in with a sample diameter of $\frac{1}{4}$ in. The anvils are tapered to facilitate calculation of pressure and to allow the thermocouple to be placed close to the sample. The angle of taper depends on the type of experiment being done. Dachille and Roy (1963) suggest 150° to 160°, which is sufficiently wide to prevent errors in the calculation of pressure due to extrusion of material. Such extrusion represents one of the main difficulties in interpreting results from opposed anvil equipment, for pressures are not uniform if material is extruded. In addition, the shearing stress of the sample itself may add a further complication. Unfortunately extrusion on anvil surfaces may not be uniform and the magnitude of this effect is largely unknown. However, the problem may be minimized by keeping the contacting anvil surfaces as small as possible and by using a thin sample. Pressures are calculated from the diameters of piston and ram and making appropriate corrections.

In order to maintain pressure and eliminate shearing stresses, the tapered anvils must meet rigid tolerances and be remachined after use. Despite such precautions, failure of anvils is common and at high pressures may have to be replaced after each run. The nature of this failure varies with the type of material used (Dachille and Roy 1963).

Details of the sample holder and recording thermocouple are shown in Fig. 7.1(c). The holder consists of a thin nickel ring surrounded by pure platinum foil. The ring is $\frac{1}{4}$ in o.d. and about $\frac{1}{8}$ in i.d. The sample is contained within this ring as a thin wafer. Studies of the Bi I–II transition across sample wafers of this type have shown that pressure distribution varies from low at the periphery to high at the centre, and is dependent on the diameter to thickness ratio of the sample; large ratios of the order of 14:1 or higher produce minimal pressure variation (Montgomery, Stromberg, Jura, G. H., and Jura, O. 1962; Myers, Duchille, and Roy 1962). Thus, in a sample assembly using a 20-ton ram, such as that shown in Fig. 7.1(c), the size of the sample is limited to 6–15 mg.

Working with such small quantities is difficult but in larger rams, in which the cross-sectional diameter of the anvils is increased, samples of up to 1 g may be used.

The platinum foil round the ring and in contact with the sample forms one junction of the two recording thermocouples; the others are loops of $Pt_{90}Rh_{10}$, or other suitable wire, attached to the foil. These thermocouples are led from the furnace to a suitable recorder. The problems of measuring of temperatures at high pressures are discussed later in this chapter.

From this brief description of the opposed-anvil apparatus, it is apparent that its principal limitations are the inability to determine pressures precisely, the small sample size, and the temperature limitations imposed by the rupture strengths of the pistons. Its major advantages are its simplicity and the accuracy of temperature measurement compared with internally heated gas media equipment. Although many syntheses of high-pressure minerals were originally done in opposed-anvil devices and attempts have been made to overcome some of the inherent difficulties (cf. Dachille and Roy 1963), most recent mineralogical and petrological high-pressure research has been done with piston–cylinder devices.

The piston–cylinder apparatus

Piston–cylinder equipment has two major advantages over opposed anvil apparatus: higher precision in pressure measurement and the ability to attain higher temperatures because of the internal furnace. The advantage of the piston–cylinder design over opposed anvils is that cylinders withstand much greater compressive stresses if they are tapered; the wider the angle of taper the greater the compression attainable (Bridgman 1952).

The piston-cylinder apparatus, designed by Boyd and England (1960a) and shown in Fig. 7.2, is a modification of an internally heated pressure system described by Coes (1955) for a temperature of 1000 °C at 45 kbar, and a later design of Hall (1958) for 2000 °C at the same pressure. This design can be used up to about 50 kbar at temperatures up to 2200 °C, and represents an improvement on the previous designs in the technique of introducing thermocouples into the sample chamber. The pressure limitation is imposed by the compressive strength of the unsupported part of the piston extending beyond the cylinder. By supporting this with a potassium bromide supporting stage, Boyd (1962) has attained pressures as high as 100 kbar.

156   *Solid-media apparatus*

The main components of the apparatus (Fig. 7.2) are the pressure-generating equipment, consisting of two hydraulic rams and a piston; the tungsten carbide pressure-vessel and steel supporting rings; and the furnace within the vessel. The pressure–vessel and supporting rings are mounted on a steel bridge. Water is circulated round the outside of the vessel to prevent thermal expansion of the hardened steel supporting ring relative to the carbide vessel.

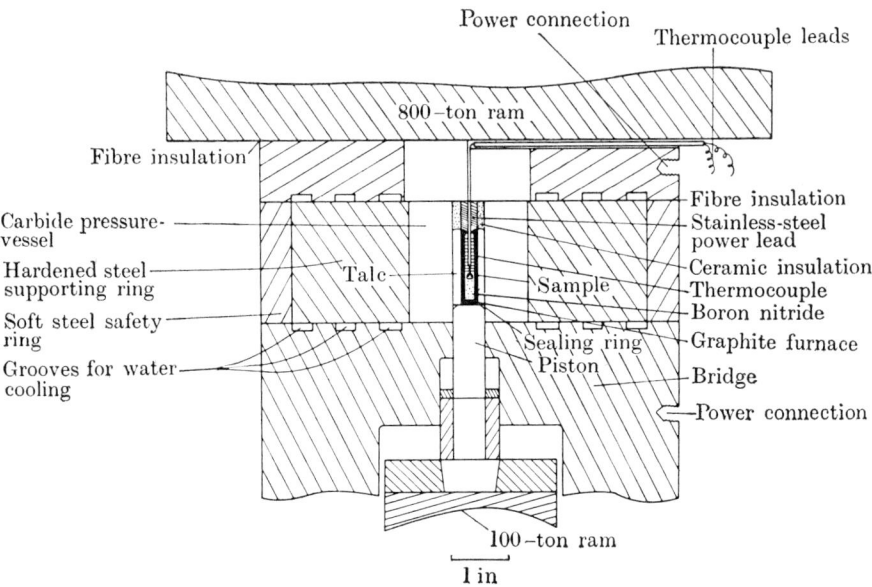

Fig. 7.2. Piston–cylinder apparatus. (After Boyd and England 1960a.)

Pressure is supplied by two hydraulic rams operated by suitable pumps. The larger one is used to end-load the vessel; the smaller ram applies pressure to a $\frac{1}{2}$-in diameter tungsten carbide piston 1 in long which acts on the sample and furnace assembly (Fig. 7.2). The bore of the pressure-vessel is 0·0005 to 0·001 in larger than the piston, which is attached to the ram with a steel casing to prevent its chipping. Boyd and England (1960a, p. 742) report that pistons of this design last indefinitely below 30 kbar and for up to 40 runs at higher pressures. Pressure is determined by measurement of the oil pressure in the press, using a conventional gauge, and from a knowledge of the cross-sectional areas of the piston and ram. Various corrections must be made to this computed pressure.

The carbide core of the vessel is made from Carbaloy† grade 883 as a cylinder approximately $2\frac{1}{2}$ in o.d., $\frac{1}{2}$ in i.d., and 2 in long tapered on the outside at a 1° angle. This is surrounded by a 2-in diameter supporting ring made from A1S1 E4340 steel, hardened to Rockwell C44–46. The bore of the ring is stretched by about 1 per cent after hardening. A safety ring of mild steel, approximately $\frac{1}{2}$ in in diamter, surrounds the support ring. The life of the pressure vessel is about 60 runs at high temperatures. When chipping and lateral cracks in the bore necessitate its replacement, this is done by pressing out the core from the support ring and inserting a new one.

The pressure-vessel and support ring assembly are separated from the large ram by a steel ring with a Carbaloy 883 inner ring (Fig. 7.2). Annealed high-pressure tubing, containing the insulated recording thermocouple passes through the centre of the Carbaloy ring. This tubing also serves as one of the power leads for the furnace. Insulation between the steel ring and pressure-vessel assembly and the steel ring and ram is provided by thin sheets of Mylar‡ or other suitable material.

The furnace and sample assembly used in the apparatus of Boyd and England is shown in detail in Fig. 7.3. The furnace is $\frac{1}{4}$ in o.d., $\frac{1}{8}$ in i.d., and $1\frac{1}{8}$ in long. The sample, as a compressed pellet in a platinum capsule 0·096 in in diameter and 0·100 to 0·125 in long, fits into the centre of the furnace (Fig. 7.3). The furnace consists of a graphite tube surrounded by a talc sleeve; the latter acts as an insulator and a pressure transmitter. A thin lead foil separates the furnace from the Carbaloy vessel. Boron nitride and high-temperature porcelain are inserted within the furnace; the high-temperature porcelain serves to separate the platinum capsule and thermocouple tip from the boron nitride, which is soluble in platinum.

A base plug at the upper end of the furnace (Fig. 7.2) is shown in detail in Fig. 7.3. This plug is 0·495 in diameter and 0·500 in long. It consists of the annealed high-pressure tubing power lead through which an insulated thermocouple is held in place by friction. At the lower end of the base plug in contact with the furnace, a ring of fired pyrophyllite encloses the steel tubing, and at the upper end a steel ring hardened to Rockwell C-55 reduces the stress at the edge of the bore of the plug. The entire plug is surrounded by a thin washer of unfired pyrophyllite to prevent chipping of the bore when pressure is released. For consistent results, the maximum tolerances within the base plug assembly should not exceed 0·001 in.

† Trade mark of the General Electric Co.
‡ Trade mark of Arthur Brank and Co. Inc., Allanton, Mass.

The power and thermocouple leads pass out of the assembly through a groove between the large ram and upper bridge (Fig. 7.2). A 220-volt 50-A transformer fed through a 5-kW transformer provides the power. Temperatures may be controlled to within ± 5 °C at 1700 °C by adding a small voltage, in phase, through a second transformer. At the end of a run, quenching is achieved by shutting off the power to the furnace, which has a low thermal inertia and cools the sample to less than 500 °C n about 5 s.

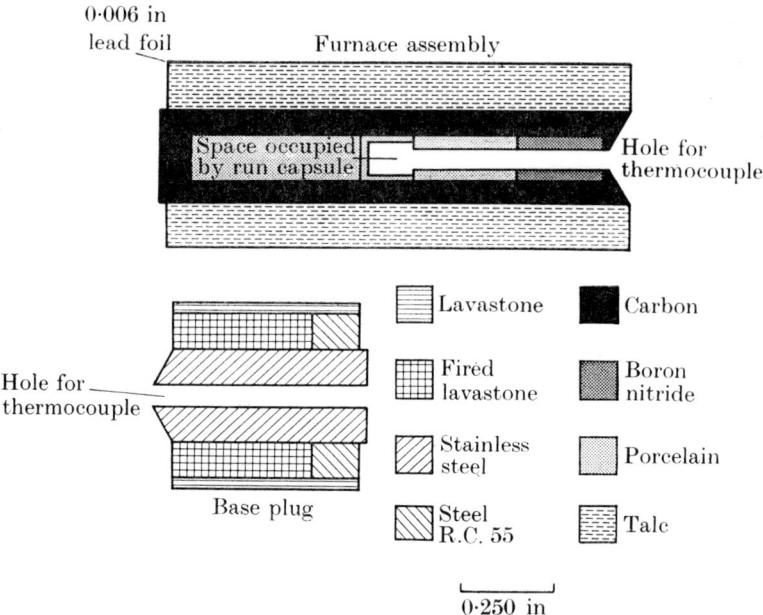

FIG. 7.3. Furnace assembly for piston–cylinder apparatus. (After Boyd and England 1960a.)

The piston–cylinder apparatus may be used either for dry or for hydrothermal experiments. In the former, the sample is enclosed in an unwelded platinum capsule or foil, whereas for hydrostatic pressures a welded capsule containing charge and water may be used. Provided that the apparatus has been carefully constructed and maintained, it is both simpler to use and less expensive than the internally heated pressure-vessels described in Chapter 6. In addition, it extends the permissable pressure range fivefold. However, in common with all solid-media devices, problems arise with calibration of both pressures and temperatures.

Other solid-media apparatus

A number of other types of solid-media apparatus, chiefly for research at pressures higher than are attainable by the piston–cylinder equipment, have been used in geological research. The equipment becomes increasingly complex and expensive as its pressure capabilities increase, and consequently it is not widely available. Two examples will be considered here, the 'belt' apparatus and the tetrahedral anvil apparatus; both were used for the first successful synthesis of diamond, and initially described by Hall (1958, 1959).

The 'belt' apparatus. The 'belt' apparatus may be considered as a hybrid of the piston–cylinder and opposed anvil devices in that it

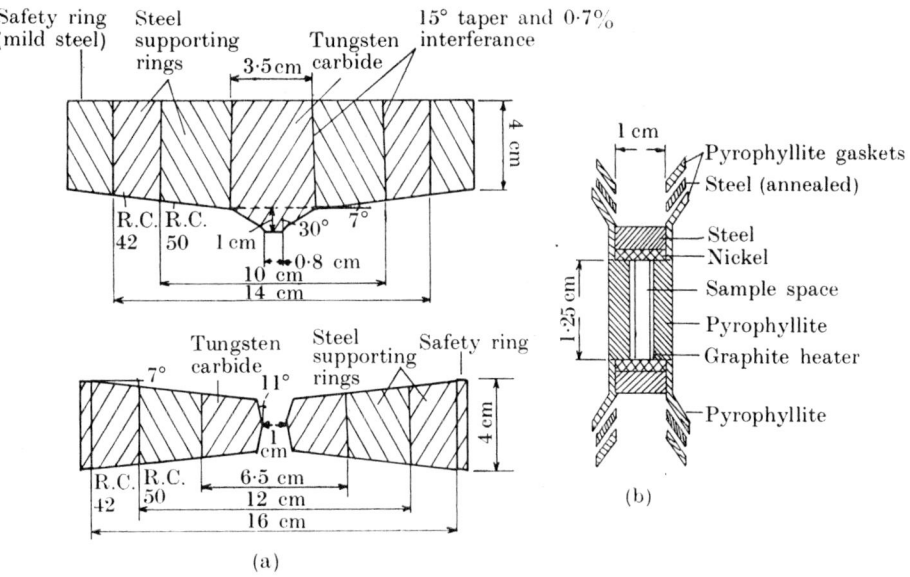

FIG. 7.4. Belt apparatus: (a) principal components; (b) sample cell and gaskets. (After Hall 1959.)

incorporates massive support for the pistons (as in the piston–cylinder apparatus) with a small central die for the samples (as in the opposed anvil apparatus). This equipment, illustrated in Fig. 7.4 from Hall's (1959) design, is capable of attaining pressures of 100 kbar at 2000 °C for runs of several hours' duration. It has the additional advantage over opposed anvil devices of accommodating large-sized samples.

Bradley (1969) describes the 'belt' apparatus in some detail; the following is a condensed version of his account. The sizes of the components of the apparatus are shown in Fig. 7.4. It is completely symmetrical about a horizontal plane through its centre and therefore only the upper piston is shown (Fig. 7.4(a)); the lower one is exactly the same. The basic principle of this equipment is that two opposing vertical pistons compress a sample which is restricted by two horizontal dies. The sample assembly (Fig. 7.4(b)) consists of conical gaskets of pyrophyllite protected by annealed steel cones compressed by the vertical pistons. The chief advantage of the conical pyrophyllite gaskets over non-coned gaskets is that compression is reduced by a factor which is a function of the sine of half the angle of the cone during vertical compression. Thus, by coning the pyrophyllite, considerably higher pressures may be attained before the pyrophyllite extrudes to its limiting thickness, as is the case for opposed anvil devices.

The tapered pistons are of tungsten carbide surrounded by hardened steel supporting rings and safety rings (Fig. 7.4(a)); this arrangement is similar to that for the piston–cylinder technique. The lower portion of the piston (about 1 cm) is tapered to a half-angle of 30° to the vertical; the supporting rings, tapered at 7°, fit exactly with the supporting rings of the die (Fig. 7.4(a)). The dies also consist of tungsten carbide tips, surrounded by steel supporting rings. The taper on the tungsten carbide tips is 11° and is such that the vertical tapered pistons may move freely into it.

The furnace and sample assembly, shown in Fig. 7.4(b), provides for samples approximately 1·25 cm long by 1 cm in diameter. The sample cell consists of a graphite cylinder, as furnace, which is surrounded by a core of pyrophyllite with nickel and steel plugs at the top and bottom. Thermocouples may be placed within the sample chamber by drilling holes through the thicker portions of the gaskets. Insulated leads for thermocouples and power supply pass between one of the die and piston assemblages.

Several modifications of the 'belt' apparatus have been made (cf. Bradley, 1969). With these, pressures up to 200 kbar may be attained by sacrificing the size of the sample cell.

The tetrahedral anvil apparatus. The tetrahedral anvil is a modification of the opposed anvil apparatus but has the advantage of larger sample capacity and more precise pressure determination. Its principal disadvantages are its size, cost, and complexity. Pressures in excess of 70 kbar at 2000 °C are possible. This device, shown schematically in Fig. 7.5, has four tungsten carbide anvils with triangular tips driven by hydraulic rams. The pressure is exerted along lines perpendicular to the faces of a pyrophyllite tetrahedron containing the sample furnace. The tetrahedron is larger than the cavity between the anvils. On advancing the pistons, the pyrophyllite is compressed and serves as the pressure medium as well as thermal and electrical insulator. Temperature is measured by a thermocouple placed within the tetrahedron close to the sample. This design, originally described by Hall(1958), has been further

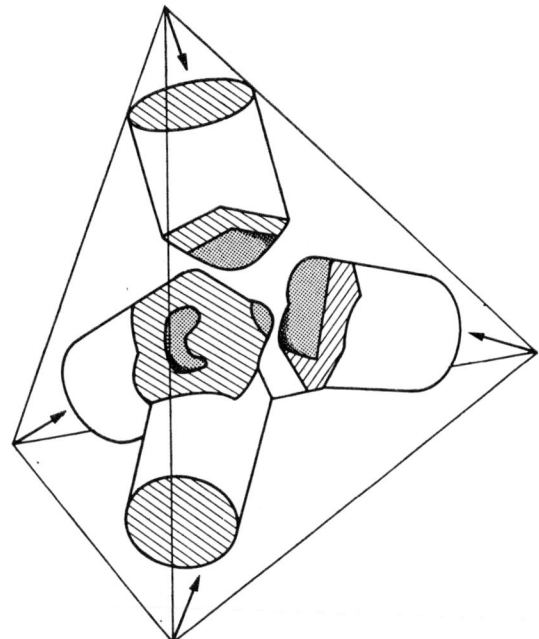

FIG. 7.5. Schematic diagram of the principle of the tetrahedral anvil. (After Hall 1958.)

developed by Liander and Lundblad (1960) to give a multi-anvil device with a cubic geometry with six pistons.

In Hall's original design, all four pistons were moved by individual hydraulic presses. In a later design (Lloyd, Hutton, and Johnson 1959) only one ram was used to activate the upper piston, the remainder being rigidly supported.

For further details of 'belt' and tetrahedral anvil equipment as well as other solid-media apparatus for high pressures, the reader is referred to Bradley (1969), Wyllie (1966), MacChesney and Rosenberg (1970), and Manning and Labrow (1971). Manning and Labrow also discuss the design and operation of auxilliary equipment, such as reciprocating

162 Solid-media apparatus

pumps for operating hydraulic rams, in considerable detail. Other accessories are the same as, or similar to, those described in previous chapters for equipment operating at lower pressures.

Calibration of temperature and pressure in solid-media apparatus

General

Calibrations and determinations of both pressures and temperatures present the major uncertainties in the results obtained from solid-media apparatus. The magnitude of both pressure and temperature errors depends on the type of equipment used and on the pressure and temperature ranges involved. Of the four varieties of apparatus described here, the piston–cylinder has been used for most petrological work and, since the description of Boyd and England (1960a) was first published, a large number of investigations of pressure and temperature calibrations have been made (cf. Boyd and England 1960a, b; Newton and Smith 1967; Boettcher and Wyllie 1968a; Kitahara and Kennedy 1964; Green, Ringwood, and Mayor 1966; Boyd, Bell, England, and Gilbert 1967; Getting and Kennedy 1970; Mao and Bell 1970; Williams and Kennedy 1969, 1970; Johannes *et al.* 1971; Bell and Williams 1971; Bell *et al.* 1971). These have resulted in considerable advances of our understanding of the calibration of both pressure and temperature in this equipment. The present discussion refers specifically to the piston–cylinder apparatus, although the calibration difficulties are similar in other solid-media equipment.

Errors in the calculated pressure are due primarily to friction caused by shearing of the pressure medium and mechanical friction between piston and vessel. In theory, corrections for these errors may be made by calibrating the apparatus at known pressure transitions. In practice, most of the suitable transitions are known only at room temperature and the effects of higher temperatures are unknown, Further calibration problems arise with the geometry and method of operation, the type of lubricant (if any), the piston and sample-cell material, etc. As a result of many attempts to calibrate the piston–cylinder apparatus, a large amount of data has been gathered, some of which appears to be contradictory and unsystematic. However some attempt at inter-laboratory calibration with different designs and materials is now being made.

Of equal importance is the calibration of temperatures in piston–cylinder equipment. The main problems in temperature calibration are the effect of pressure on the e.m.f.s of various types of thermocouples, errors due to temperature gradients, and errors caused by contamination. Progress is being made on these by improvements in thermocouple insertion and by using thermocouple wires whose e.m.f. is not pressure-dependent.

Pressure calibration

The non-hydrostatic† pressure distribution in piston–cylinder apparatus is caused by friction between the piston and cylinder walls and by differential compressibilities and strengths of materials used in the pressure cell. Although in theory it may appear simple to make a correction for both these factors in practice this is extremely difficult because of the operating conditions, the geometry of the apparatus, the materials used in the pistons, the 'smoothness' of piston and cylinder walls, etc. Estimates of the size of pressure correction range from zero to as high as 25 per cent (Johannes, Bell, Mao, Boettcher Chipman, Hays, Newton, and Seifert 1971). Further problems arise in the choice of reaction for pressure, calibration which in turn may be affected by the condition of the starting-materials, etc.

Pressure errors caused by frictional affects between cylinder and piston walls vary depending on whether compression (piston-in) or decompression (piston-out) techniques are employed. For a given transition at fixed temperature, the piston-in technique produces a higher pressure for the transition than the piston-out technique, resulting in a pressure hysteresis which may be as large as 9 per cent of the true transition pressure. This method of operation will thus produce either a positive or negative correction from the calculated value. The size of the hysteresis pressure loop is not constant for all piston–cylinder equipment but depends on its geometry and on whether a piston of a single material or a piston with both carbide and steel parts is used. These factors have resulted in a variety of values for the correction being reported by investigators, using equipment of different designs. Johannes *et al.* (1971, p. 25) list various optimum conditions in which the maximum and minimum pressures exceed true pressures using piston-in and piston-out techniques respectively.

Errors in pressure calibration caused by differences in the properties of materials used in the pressure cell are more difficult to detect and measure owing to the large number of components in the cell (often seven or more). Some of these are much more compressible than others and

† Non-hydrostatic pressures are those in which the applied force per unit area is not perpendicular to the area.

may behave differently under piston-in or piston-out conditions. The contacts between various types of material in a cell under pressure and the piston stroking technique, may cause pressure errors to cancel one another out or be additive. Bell et al. (1971) give a number of possible idealized examples of the types of errors which may arise from combinations of various materials in a pressure cell.

Calibration problems are compounded by lack of knowledge of the effect of temperature on the transition pressures of many common calibrants. In their early work with piston–cylinder apparatus, Boyd and England (1960a, b) determined frictionally-produced pressure corrections using the BiI–BiII and TlI–TlII transitions at 25 °C giving a pressure correction of −13 per cent. For higher temperatures, the quartz–coesite transition gave a correction of −8 per cent (Boyd and England 1960b). Other workers (cf. Newton and Smith 1967; Boettcher and Wyllie 1968a) used calibrants such as LiCl whose transitions had previously been determined in gas-media apparatus in which the pressures much more closely approach ideal hydrostatic conditions. Some results may be in error because of such factors as the type of lubricant used (Boettcher and Wyllie 1968a). In some cases the uncertainties in the correction were almost as large as the correction itself.

The pressure–temperature breakdown curve for the reaction

$$\text{albite} \rightarrow \text{jadeite} + \text{quartz}$$

has been examined by many investigators either directly (Birch and Le Comte 1960; Bell and Rooseboom 1965; Newton and Smith 1967; Newton and Kennedy 1968; Boettcher and Wyllie 1968b) or from thermodynamic calculations and extrapolations (Adams 1953; Kelley, Todd, Orr, King, and Bonnichsen 1953; Hlabse and Kleppa 1968). Although the pressure of this reaction at a given temperature is not exactly known and the ranges of pressure transition with temperature are not particularly low, it provides a suitable pressure calibrant, at about 10 kbar at 600 °C, which can easily be reversed hydrothermally under completely hydrostatic pressure conditions.

In an attempt to clarify some of the difficulties of pressure calibration in piston–cylinder equipment and to give some indication of the magnitudes of pressure calibration errors in different laboratories with equipment of various geometries and sample-cell designs, the reaction albite → jadeite + quartz has been determined at 600 °C by six sets of investigators. The results and their implications are discussed by Johannes et al. (1971) and Bell, Mao, and England (1971). All investigators used the same starting-materials enclosed in sealed capsules with 20 to 50 weight per cent water added. Average 'corrected' pressure values of the transition ranged from 15·7 kbar to 16·8 kbar, with estimated errors ranging from about ±0·5 kbar to ±1·5 kbar. The average value of all measurements was 16·3 kbar, with the error range from all investigators falling within this value. Johannes et al. (1971) and Bell et al. (1971) consider this agreement to be good in view of the different experimental procedures and designs used by each group. However, as Bell et al. (1971) point out, this pressure value can be considered as only relative and systematic errors due to inhomogeneous stress distributions may still be present in these results.

Other pressure calibrants for piston–cylinder apparatus and a discussion of problems of calibration are given by Bell and Williams (1971).

Temperature calibration

The measurement and calibration of thermocouples in piston–cylinder and other solid media apparatus is subject to both random and systematic sources of error. Both types of error are dependent on the pressures and temperatures used but tend to become more pronounced at temperatures above 1000 °C and high pressures. Random errors include temperature gradients (these are much smaller in solid-media apparatus than in high-pressure internally heated gas equipment), temperature differences resulting from differential pressures on the thermocouple produced by the high strength of its insulating ceramic, and temperature variation and drift due to contamination of the wires. Systematic errors also include contamination from the materials of the sample cell, contamination of one thermocouple wire by diffusion of material from the other wire under pressure, and tension of the thermocouple wires under pressure. Errors due to both pressure affects and contamination will result in erroneous e.m.f.s; although those caused primarily by contamination will produce increasing deviations in e.m.f. with increasing contamination.

During the period 1960–70, when much high-pressure work was done with piston–cylinder apparatus, most investigators used Pt–Rh thermocouples for temperature measurement above 1000 °C. Although many workers (cf. Boyd and England 1963; Williams and Kennedy 1969) inferred that Pt–Rh thermocouples were in error under high-pressure conditions, it was not until the investigations of Getting and Kennedy (1970), Mao and Bell (1970) and Mao, Bell, and England (1971) that any attempt was made to assess the causes and magnitude of these errors and to suggest alternative thermocouples.

Getting and Kennedy (1970) determined temperature errors in piston–cylinder apparatus up to 1000 °C at 33 kbar, and Mao and Bell (1970) extended this work to 1700 °C at 40 kbar. Mao and Bell determined inherent random uncertainties and corrections for systematic errors for both the commonly used $Pt_{90}Rh_{10}$ thermocouples and for thermocouples of $W_{90}Re_3$–$W_{75}Re_{25}$. Mao et al. (1971) extended this investigation by measuring additional sources of error caused by tension on the thermocouple wires and thermocouple drift due to contamination of the wires by various materials of the sample cell.

Mao and Bell (1970) discovered that the W–Re thermocouples are superior to the Pt–Rh thermocouples in both their inherent lower random errors and in chemical inertness. They found that Pt–Rh thermocouples become particularly sensitive to contamination when used for prolonged periods above 1600 °C in the absence of strong insulating materials. Mao and Bell (1970) observed that random thermocouple errors could be reduced by keeping Pyrex glass in contact with the thermocouple wires, and by using weak materials such as talc, boron nitride, and Pyrex glass for the low-temperature parts of the pressure cell.

For their experiments on tension and contamination as contributors to temperature errors in piston–cylinder apparatus, Mao et al. (1971) used the same thermocouple materials used by Mao and Bell (1970). The thermocouple wires were carefully calibrated and subjected to various pressures using different pressure cell designs and thermocouple arrangements. Changes in the e.m.f.s of both types of thermocouple were determined at temperatures up to 800 °C under tensional pressures up to 5·8 kbar. These experiments confirmed the superiority of the W–Re thermocouples over the Pt–Rh ones; the latter showed larger negative e.m.f. changes at high temperature. A further major advantage of the W–Re thermocouple is that the tensile strengths of each wire are both higher than those of Pt–Rh wires and are unaffected by temperature. In contrast, the tensile strengths of Pt and Pt–Rh alloy wires decrease with increasing temperature. Mao et al. (1971) conclude that compression on one thermocouple wire combined with tension on the other could lead to errors of 6–10 °C at 1000 °C, and that asymmetrical hysteresis loops during cycling of temperature and pressure are developed which are not directly related to pressure hysteresis loops on the sample cell.

In their experiments on the effects of contamination on the two thermocouples, Mao et al. (1971) subjected the wires, encased in combinations of high-grade alumina and Pyrex glass, to temperatures up to 1900 °C at 20 kbar. The results showed that serious contamination begins above 1300 °C and is nearly constant up to 1600 °C, where drift in e.m.f. caused by contamination can amount to 30 °C after 3 h. These experiments indicated that thermocouple drift was greater in Pt–Rh wires than W–Re ones, and that in the latter, the $W_{97}Re_3$ wire is more prone to contamination at high temperatures. They also showed that drift caused by contamination is minimized if the wires are completely covered by high-grade alumina.

Mao et al. (1971) have attempted to assess corrections and uncertainties in thermocouple measurements from all possible sources. They suggest that for temperatures below 1700 °C the optimum conditions are a $(W_{97}Re_3)$–$(W_{75}Re_{25})$ thermocouple insulated with Pyrex glass. This produced an uncertainty in temperature from all sources of approximately ± 0.5 per cent $+ 0.2$ per cent per 10 kbar. Above 1700 °C W–Re thermocouples should be insulated with high purity alumina giving an approximate temperature uncertainty of ± 0.7 per cent $+ 0.2$ per cent per 10 kbar. Both uncertainties are for runs of several hours' duration.

Operation, maintenance, and safety precautions

The operation of solid-media apparatus becomes more complex as the geometry and numbers of stages of compression increase. The opposed single-stage anvil apparatus is much simpler to operate than the tetrahedral anvil or other multi-anvil devices with many pistons. Successful operation depends on extremely careful building and maintenance of the equipment. Obviously the heads of pistons, tapered cylinders, etc., must be kept at exact tolerances if the equipment is to operate properly and if large errors in pressure and temperature measurement are to be avoided. All pressure-contact surfaces must be scrupulously clean and kept at a high polish.

As in high-pressure gas-media apparatus, certain parts of solid-media apparatus, both in the sample cell and in the pressurizing assembly, tend to wear and must be renewed. Details of the expected lives of various parts are given in the references in this chapter. Defective or worn parts in the pressure systems of gas-media apparatus are usually readily detected by pressure leaks, but worn parts in solid-media equipment do not have so dramatic an effect and the more readily wearable parts should be examined after each run for fractures or other signs of wear.

Routine operation of hydraulic pumps, etc., is described in many texts (cf. Manning and Labrow 1971) and information on operating the equipment can be found in the references cited. The simpler types of solid-media apparatus are easier to operate than gas-media equipment but their maintenance is likely to be more difficult owing to the nature of the pressure media and to uncertainties in pressure and temperature calibrations.

As in all high-pressure equipment, safety measures are of the utmost importance. Although not perhaps as dangerous as gas-media apparatus, despite its higher pressures, solid media apparatus must be well shielded. In addition to the dangers of operating the equipment, there are also dangers in its construction and maintenance. For example in the pressing of the tightly fitting rings of the vessel in the piston–cylinder apparatus (Fig. 7.2), uneven pressing or differential contraction can result in explosive shattering of the ring even after it has been removed from the press. After assembly, the apparatus must have a sufficiently rigid framework to support the high pressures involved.

References

ADAMS, L. H. (1953). *Am. J. Sci.* **251**, 299.
BELL, P. M. (1967). *J. geophys. Res.* **72**, 666.
—— MAO, H. K. and ENGLAND, J. L. (1971). *Carnegie Inst. Wash. Year Bk.* **70**, 277.
—— and ROOSEBOOM, E. H. (1965). *Carnegie Inst. Wash. Year Bk.* **64**, 139.
—— and WILLIAMS, D. W. (1971). In *Research techniques for high pressures and high temperatures* (ed. G. C. Ulmer), Springer–Verlag, New York, p. 195.
BIRCH, F. and LeCOMTE, P. (1960). *Am. J. Sci.* **258**, 209.
BOETTCHER, A. L. and WYLLIE, P. J. (1968a). *Contr. Miner. Petr.* **17**, 224.
—— —— (1968b). *Geochim. cosmochim. Acta.* **32**, 999.
BOYD, F. R. (1962). In *Modern very high pressure techniques* (ed. R. H. Wentorf). Butterworths, Washington, p. 151.
—— BELL, P. M., ENGLAND, J. L. and GILBERT, M. G. (1967). *Carnegie Inst. Wash. Year Bk.* **66**, 410.
—— and ENGLAND, J. L. (1960a). *J. geophys. Res.* **65**, 741.
—— —— (1960b). *J. geophys. Res.* **65**, 749.
—— —— (1963). *J. geophys. Res.* **68**, 311.
BRADLEY, C. C. (1969). *High pressure methods in solid state research*. Butterworths, London, p. 176.
BRIDGMAN, P. W. (1949). *Proc. Am. Acad. Arts Sci.* **77**, 117.
—— (1952). *The physics of high pressure*. Bell, London.
—— (1964). *Collected experimental papers*, vols. 1–7. Harvard University Press, Cambridge, Mass.

COES, L. (1955). *J. Am. ceram. Soc.* **38**, 298.
DACHILLE, F. and ROY, R. (1960). *Am. J. Sci.* **258**, 225.
—— (1962). In *Modern very high pressure techniques* (ed. R. H. Wentorf). Butterworths, Washington D.C., p. 163.
—— —— (1963). In *The physics and chemistry of high pressures*, Society of Chemical Industry Symposium, London, p. 77.
GETTING, I. C. and KENNEDY, G. C. (1970). *J. appl. Phys.* **41**, 4552.
GREEN, D. H., RINGWOOD, A. E. and MAJOR, A. (1966). *J. geophys. Res.* **71**, 3589.
GRIGGS, D. T. and KENNEDY, G. C. (1956). *Am. J. Sci.* **254**, 722.
HALL, H. T. (1958). *Rev. scient. Instrum.* **29**, 267.
—— (1959). *Rev. scient. Instrum.* **31**, 125.
HLABSE, T. and KLEPPA, O. J. (1968). *Am. Miner.* **53**, 1281.
JOHANNES, W., BELL, P. M., MAO, H. K., BOETTCHER, A. L., CHIPMAN, D. W., HAYS, J. F., NEWTON, R. C. and SEIFERT, F. (1971). *Contr. Miner. Petrol.* **32**, 24.
KELLEY, K. K., TODD, S. S., ORR, P. L., KING, E. G. and BONNICHSEN, J. R. (1953). *Rep. Invest. U.S. Bur. Mines.* 4955.
KITAHARA, S. and KENNEDY, G. C. (1964). *J. geophys. Res.* **69**, 5395.
LIANDER, H. and LUNDBLAD, F. (1960). *Arkio. Kemi* **16**, 139.
LLOYD, E. C., HUTTON, V. O. and JOHNSON, D. P. (1959). *Bur. Stand. J. Res.* C **63**, 59.
MACCHESNEY, J. B. and ROSENBERG, P. E. (1970). In *Phase diagrams Materials science and technology*, vol. 1 (ed. A. M. Alper). Academic Press, London, p. 113.
MANNING, W. R. D. and LABROW, S. (1971). *High pressure engineering*. The Chemical Rubber Co., Cleveland, Ohio, p. 369.
MAO, H. K. and BELL, P. M. (1970). *Carnegie Inst. Wash. Year Bk.* **69**, 207.
—— BELL, P. M. and ENGLAND, J. L. (1971). *Carnegie Inst. Washington Year Bk.* **70**, 281.
MONTGOMERY, P., STROMBERG, H., JURA, G. H. and JURA, O. (1962). *Am. Soc. Mech. Eng. Publ.* 64–Wa–308, 1.
MYERS, M. B., DUCHILLE, F. and ROY, R. (1962). *Am. ceram. Soc. Bull.* **41**, 225.
NEWTON, M. S. and KENNEDY, G. C. (1968). *Am. J. Sci.* **266**, 728.
NEWTON, R. C. and SMITH, J. V. (1967). *J. Geol.* **75**, 268.
PATERSON, M. S. (1970). In *Mechanical behaviour of materials under pressure* (ed. A. Pugh), Elsevier, Barking, Essex, p. 191.
SIMMONS, G. (1968). *J. geol. Education* **16**, 21.
VODAR, B. and KIEFFER, J. (1970). In *Mechanical behaviour of materials under pressure* (ed. A. Pugh), Elsevier, Barking, Essex, p. 1.
WILLIAMS, D. W. and KENNEDY, G. C. (1969). *J. geophys. Res.* **74**, 4359.
—— —— (1970). *Am. J. Sci.* **269**, 481.
WYLLIE, P. J. (1963). *J. geophys. Res.* **68**, 4611.
—— (1966). In *Methods and techniques in geophysics*, (ed. S. K. Runcorn) Interscience, New York, p. 33.

8. Control of partial pressures of volatile components at high total pressures

Introduction

IN Chapters 5, 6, and 7 experimental techniques for high load and water vapour pressures have been described. The techniques of mixing gases to provide partial pressures of oxygen, CO_2, and other gases at a total pressure of about 1 atmosphere were outlined in Chapter 4. This chapter is concerned with experiments at high pressures in which the partial pressures of volatiles other than water are controlled, and experiments in which the partial volatile pressure is less than the total pressure. Although they are not considered here, the P–V–T relations of H_2O, CO_2, and other geologically important volatiles are fundamental in that they provide the basic thermodynamic data for experiments with controlled pressures of volatile components.

The importance of volatiles, particularly oxygen in reduction and oxidation reactions and CO_2 in metamorphic processes, had long been recognized, but it was not until the late 1950s that laboratory techniques were developed in which $p(O_2)$, $p(CO_2)$,[†] and other partial pressures of volatile components could be controlled at total pressures greater than one atmosphere. This led to the synthesis and determination of the stability fields of a large number of important rock-forming silicates containing polyvalent ions under controlled partial pressures, an evaluation of redox reactions in metamorphism, and a better understanding of the influence of volatile (mobile) components in igneous and metamorphic processes.

After the work of Darken and Gurry (1945, 1946, 1953) on the system Fe–O and that of Muan and Osborn (cf. Muan 1958, Osborn 1959, Muan and Osborn 1956) on other silicate and oxide systems containing iron at 1 atmosphere, Eugster and his co-workers developed buffering techniques for controlling $p(O_2)$ (and hence $p(H_2)$) at total pressures up to several kilobars. Similar methods have since been developed to control $p(CO_2)$, $p(CO)$, $p(S)$, $p(F)$, etc.

[†] Partial volatile pressures are denoted by 'p', total volatile pressures by 'P'.

In the techniques described in earlier chapters, the pressures of volatiles are equal to the total (or confining) pressures. However, in nature, pressures of volatiles may be less than confining pressures owing to the fluid phase having more than one component, the pore fluid supporting only part of the rock-mass (the remainder being supported by the minerals of the rock), or the chemical potential of one or more of the volatiles being affected by some source outside the rock-system. Conditions in which pressures of volatiles are less than confining pressures are possibly more common in metamorphic than in igneous rocks, particularly during metamorphism of rocks obviously containing more than one volatile component, such as wet calcareous sediments. Greenwood (1961, Fig. 3) describes an experimental approach to this problem and gives a detailed discussion of the geological implications of pressures of volatiles becoming less than total pressures.

Solubility experiments form a very important branch of experimental petrology and are discussed briefly at the end of Chapter 5. This type of study involves the solubilities of water and other volatiles in melts as well as the solubilities of solids in liquids. These experiments provide information on a host of geological phenomena, including explosive volcanism, the origins of pegmatites, metasomatism, and the formation of ore solutions.

The buffer technique

Theory of buffers

In geological processes, chemical equilibrium is defined in terms of the total pressure, temperature, and chemical potentials of the components present in the system. For any fixed temperature and pressure, the chemical potentials of the components are defined by their equilibrium constants. The equilibrium constants of gaseous components are here expressed as fugacities and those of solid and liquid components as activities (Glasstone 1946). Many geological reactions involve solid, liquid, and gas phases. The liquid may be an aqueous solution or a melt in which the gaseous species are soluble and therefore influence the solubility of the solid phases and the crystallization of solids from the melt. The gaseous species may contain several components whose effects on solubility and crystallization must be known. Eugster and Skippen (1967) have outlined methods for studying the effects of systems containing up to four gas components and much of the following section is taken from their paper.

172   *Partial pressures of volatile components*

Partial volatile pressures, at 1 atmosphere total pressure, can be controlled by using gas mixtures, as described in Chapter 4. At high total pressures, these methods cannot be used because of the difficulty of evaluating the effect of total pressure on partial pressures, and it is necessary to employ solid assemblages to control gas fugacities. Such assemblages, known as buffers, define and keep constant a single gas fugacity, or the ratio of several gas fugacites, for a particular total pressure and temperature (Eugster and Skippen 1967, p. 497). For each buffer the gas fugacity can be calculated or calibrated as a function of temperature and total pressure of the system. Buffers have been developed for reactions involving gases in the systems O–H[†] (Eugster 1957, 1959; Eugster and Wones 1962), C–O and C–O–H (French and Eugster 1965), H–O–S and C–O–H–S (Eugster and Skippen 1967); and H–O–F (Munoz and Eugster 1969). Buffering techniques for systems containing other volatiles such as boron, nitrogen, chlorine, and phosphorus are also possible.

Three types of buffer are in common use.

(a) The oxygen buffers described by Eugster (1957), Eugster and Wones (1962), and Huebner and Sato (1970) for gases in the system O–H.
(b) The graphite buffer developed by French and Eugster (1965) for gases in the system C–O.
(c) A combined oxygen and graphite buffer described by French and Eugster (1965) for gases in the system C–O–H.

The oxygen buffers differ from the graphite buffer in that they define $f(O_2)$[‡] for total given pressures and temperatures and hence control individual oxygen fugacities. In contrast, the graphite buffer (and therefore any combined oxygen and graphite buffer) defines $f(O_2)$ for a given gas pressure and temperature only if no gases other than CO_2, CO, and O_2 are present in the system, and therefore buffers with respect to fugacity ratios.

Oxygen buffers. An oxygen buffer consists of an assemblage of solids[§] together with water which produce a fixed $f(O_2)$ for a given temperature and total pressure. The presence of water also results in a fixed $f(H_2)$ because of the thermal dissociation reaction

$$2H_2O \rightleftharpoons 2H_2 + O_2$$

[†] Chemical symbols are used here as shorthand for the elements, without implying any particular molecular species.
[‡] See Chapter 2.
[§] Solids are normally used since they have a large buffering capacity per unit volume.

Partial pressures of volatile components 173

The equilibrium constant (K_W) for this reaction is

$$K_W = \frac{f(H_2O)}{f(H_2)f(O_2)^{\frac{1}{2}}}. \tag{8.1}$$

The total gas pressure (P_{gas}) in the system O–H, assuming H_2O, H_2, and O_2 to be the only gaseous species, is given by†

$$P_{gas} = p(H_2O) + p(H_2) + p(O_2)$$
$$= \frac{f(H_2O)}{\gamma(H_2O)} + \frac{f(H_2)}{\gamma(H_2)} + \frac{f(O_2)}{\gamma(O_2)}. \tag{8.2}$$

Under geological conditions, the last term in eqn (8.2) is negligible compared with the other terms. Therefore (8.2) reduces to

$$P_{gas} = \frac{f(H_2O)}{\gamma(H_2O)} + \frac{f(H_2)}{\gamma(H_2)}. \tag{8.3}$$

In equations (8.1) and (8.3), $f(H_2O)$ and $f(H_2)$ are the only unknowns and can be solved provided the equilibrium constant (K_W) for H_2O and the fugacity coefficients of water and hydrogen are available. The former are given in JANAF tables (1960) and the latter in Holser (1954) and Shaw and Wones (1963).

On the assumption that H_2O, H_2, and O_2 are the only gas species in the system, the values of $f(H_2O)$ and $f(H_2)$ can be obtained from (8.1) and (8.3) as

$$f(H_2O) = \frac{P_{gas} K_W f^{\frac{1}{2}}(O_2) \gamma(H_2) \cdot \gamma(H_2O)}{K_W f^{\frac{1}{2}}(O_2) \gamma(H_2) + \gamma(H_2O)} \tag{8.4}$$

and

$$f(H_2) = \frac{P_{gas} \gamma(H_2) \cdot \gamma(H_2O)}{K_W f^{\frac{1}{2}}(O_2) \gamma(H_2) + \gamma(H_2O)}. \tag{8.5}$$

Eugster and Wones (1962) calculate $\log f(O_2)$ at 1 atmosphere pressure for seven solid assemblages with water for use as oxygen buffers from the equation

$$\log f(O_2) = -\frac{A}{T} + B + \frac{C(P-1)}{T} \tag{8.6}$$

where A, B, and C are constants, P is the total pressure, and T is the temperature in K. The oxygen buffers used and the reactions‡ involved are:

† See Chapter 2.
‡ Water, although necessary for all buffers, has been omitted from the equations.

QFI†

$$2\text{Fe} + \text{SiO}_2 + \text{O}_2 \rightleftharpoons \text{Fe}_2\text{SiO}_4$$
(iron) (quartz) (fayalite)

MI

$$1\tfrac{1}{2}\text{Fe} + 2\text{O}_2 \rightleftharpoons \tfrac{1}{2}\text{Fe}_3\text{O}_4$$
(iron) (magnetite)

WI

$$2\text{Fe} + \text{O}_2 \rightleftharpoons 2\text{'FeO'}(\text{Fe}_{1-x}\text{O})$$
(iron) (wüstite)

MW

$$6\text{'FeO'}(\text{Fe}_{1-x}\text{O}) + \text{O}_2 \rightleftharpoons 2\text{Fe}_3\text{O}_4$$
(wüstite) (magnetite)

QFM

$$2\text{Fe}_2\text{SiO}_4 + \text{O}_2 \rightleftharpoons 2\text{Fe}_3\text{O}_4 + 3\text{SiO}_2$$
(fayalite) (magnetite) (quartz)

NNO

$$2\text{Ni} + \text{O}_2 \rightleftharpoons 2\text{NiO}$$
(nickel) (nickel oxide)

HM

$$4\text{Fe}_3\text{O}_4 + \text{O}_2 \rightleftharpoons 6\text{Fe}_2\text{O}_3$$
(magnetite) (hematite)

Other buffers are:

CC (Ernst 1962)

$$2\text{Cu} + \tfrac{1}{2}\text{O}_2 \rightleftharpoons \text{Cu}_2\text{O}$$
(copper) (cuprite)

HB (Huebner and Sato 1970)

$$4\text{Mn}_3\text{O}_4 + \text{O}_2 \rightleftharpoons 6\text{Mn}_2\text{O}_3$$
(hausmamite) (bixbyite)

MH (Huebner and Sato 1970)

$$6\text{'MnO'}(\text{Mn}_{1-x}\text{O}) + \text{O}_2 \rightleftharpoons 2\text{Mn}_3\text{O}_4$$
(manganosite) (hausmannite)

Results of the calculations of Eugster and Wones (1962) are shown as a log $f(\text{O}_2)$–T plot in Fig. 8.1. Most geological reactions probably take place under conditions bounded by the HM and MW buffers. At a total pressure of 2000 bars the manganosite–hausmannite buffer has an $f(\text{O}_2)$ between the magnetite–hematite and nickel–nickel oxide buffers.

Because buffers may alloy with the noble metal capsules which enclose buffer and charge (see p. 184), certain capsules are unsuitable. Ag–Pd capsules cannot be used with NNO buffer and gold should not be used with CC buffer.

† Buffers are normally abbreviated according to their solid constituents.

Partial pressures of volatile components 175

The effects of total pressure on the oxygen and hydrogen fugacities obtained from oxygen buffers have been considered by Ernst (1960) and Eugster and Wones (1962). Under pressure the changes in volume of the solid buffers affect the free energies of the reactions. The non-ideal

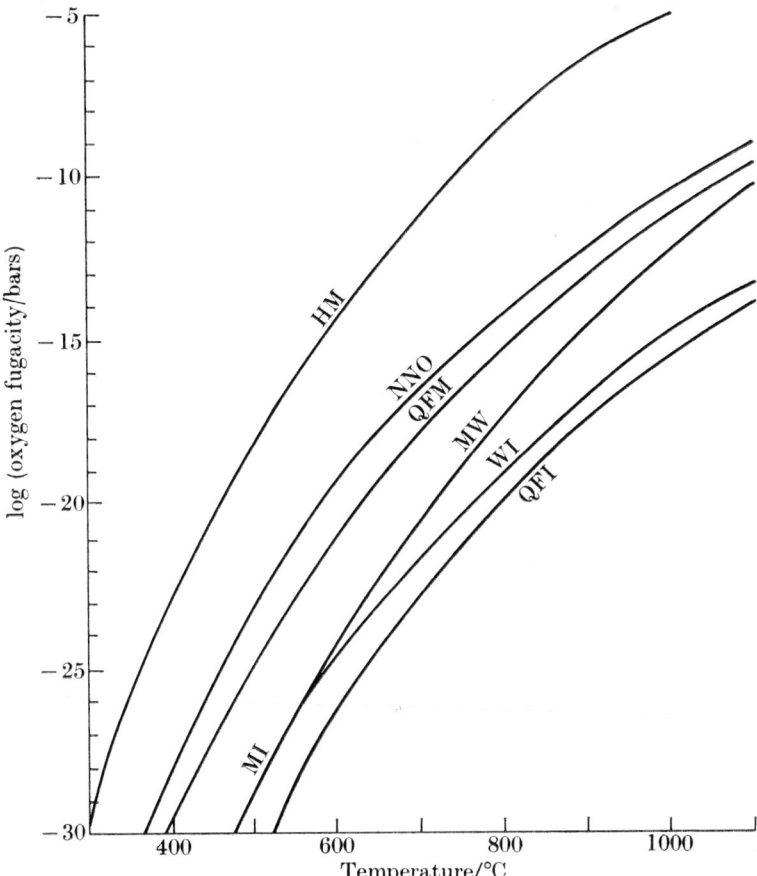

FIG. 8.1. Log (oxygen fugacity)–temperature plot for common oxygen buffers. (After Eugster and Wones 1962.)

behaviour of oxygen at high pressure will also influence the partial equilibrium oxygen pressure. The second of these factors can be eliminated by using oxygen fugacities rather than pressures, since

$$f = \gamma P, \tag{8.7}$$

where γ, the fugacity coefficient, indicates the variance of the real gas from ideal behaviour.

By definition, the fugacity of a real gas for a temperature T is

$$dG = RT\,d\ln f \tag{8.8}$$

and

$$\left(\frac{dG}{dP}\right)_T = V. \tag{8.9}$$

From (8.8) and (8.9) and using partial molar quantities,

$$\left(\frac{\partial G}{\partial P}\right)_T = \bar{V}_s = RT\left(\frac{\partial \ln f}{\partial P}\right)_T, \tag{8.10}$$

where $\bar{V}_s$ is the partial molar volume of the solid. At equilibrium, a reaction at constant temperature $d\Delta\bar{G} = 0$ (see Chapter 2, p. 18). Hence,

$$d\Delta\bar{G}_{react} = 0 = \Delta\bar{V}_s dP_t + RT\left(\frac{\partial \ln f}{\partial P}\right)_T dP_g, \tag{8.11}$$

where P_t is the pressure on the system as a whole,
P_g is the partial pressure of the gas on the system,
f_1 is the fugacity of the gas at pressure p_1.

Therefore between total pressure P_1 and P_2 the fugacity differences f_1 and f_2 are:

$$RT\ln\frac{f_2}{f_1} = -\int_{P_1}^{P_2} \Delta\bar{V}_s dP_t \tag{8.12}$$

or

$$\log\frac{f_2}{f_1} = -\frac{\Delta\bar{V}_s}{2\cdot303R} + \frac{(P_2 - P_1)}{T}. \tag{8.13}$$

In (8.13) the value of $\Delta\bar{V}_s$ over the common total pressure ranges used is virtually constant and hence the term $(\Delta\bar{V}_s/2\cdot303R)$ can be used as a correction factor for each buffer. This factor is multiplied by the total pressure in bars and divided by the temperature (in K) and added to the $f(O_2)$ value for 1 atmosphere to obtain the total pressure correction. For a total pressure of 2000 bars the $f(O_2)$ value for the common buffers increases by only about half an order of magnitude in comparison to the $f(O_2)$ at one atmosphere. Ernst (1960, pp. 14–15) gives an alternate method of calculating the effect of total pressure on $f(O_2)$ using the equilibrium constant for the solid buffer and the equilibrium constant of water.

Partial pressures of volatile components

In contrast to the small effect that total pressure has on $f(O_2)$, $f(H_2O)$ and $f(H_2)$ show large increases with increasing total pressure since one is in fact controlling hydrogen permeability. These may be calculated from eqns (8.4) and (8.5). Fig. 8.2 shows the calculated $f(H_2)$ values from

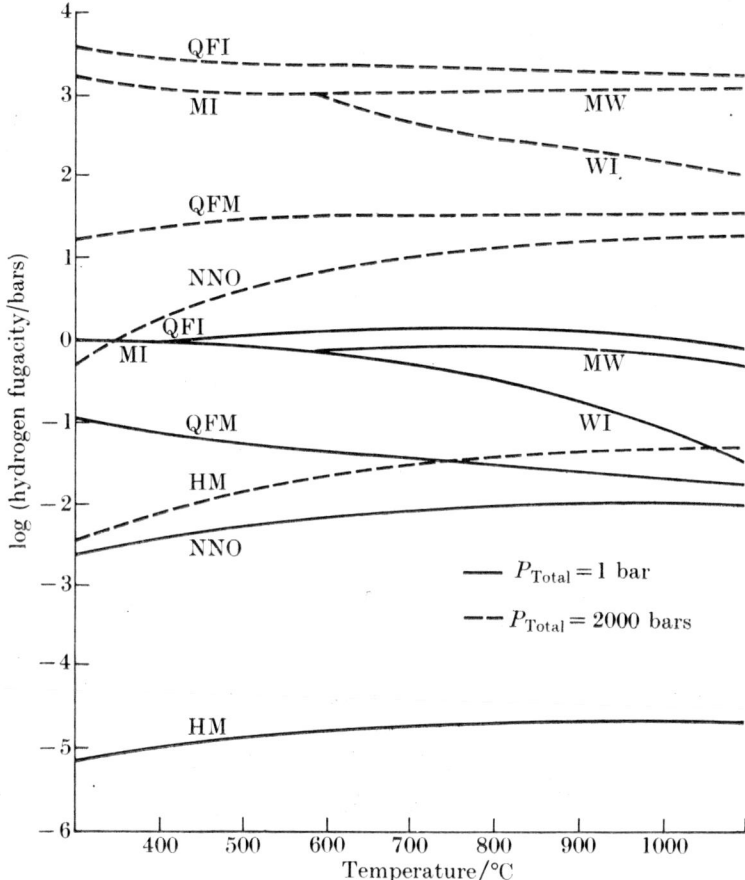

FIG. 8.2. Log (hydrogen fugacity)–temperature plot for common oxygen buffers at 1 bar and 2000 bars total pressure. (After Eugster and Wones 1962.) For abbreviations see text.

Eugster and Wones (1962) for 1 bar and 2000 bars for the common oxygen buffers. Data for $f(H_2O)$ at different total pressures are given by Eugster and Skippen (1967, p. 499), who have also shown that $f(H_2)$ can be considered as only approximate for buffer mixtures rich in H_2O and H_2 since such mixtures are non-ideal. Low H_2 values in the HM, NNO,

Partial pressures of volatile components

and QFM buffers do not, however, seriously affect $f(H_2)$. For any new buffer assemblage, calculations should be made to show the effect of total pressure on the fugacities of the gases being controlled.

Graphite buffers. The graphite buffer may be used to control CO_2, CO, and O_2 fugacities in the system C–O in which CO_2 and CO are the most abundant species. Owing to the lack of suitable noble metal tubes permeable to CO_2 and CO, experiments with a graphite buffer must be done in a $CO_2 + CO$ atmosphere as described on p. 184.

If CO_2, CO, and O_2 are the only gaseous species in the system C–O, $f(CO_2)$ and $f(CO)$ may be determined using oxygen buffers according to the reaction

$$CO + \tfrac{1}{2}O_2 \rightleftharpoons CO_2,$$

in which the value of $f(O_2)$ fixes the $CO_2:CO$ ratio. The total gas pressure is then

$$P_{gas} = p(CO_2) + p(CO) + p(O_2) \qquad (8.14(a))$$

$$= \frac{f(CO_2)}{\gamma(CO_2)} + \frac{f(CO)}{\gamma(CO)} + \frac{f(O_2)}{\gamma(O_2)} \qquad (8.14(b))$$

and

$$K = \frac{f(CO_2)}{f(CO).f(O_2)^{\frac{1}{2}}} \qquad (8.15)$$

where K is the dissociation constant for CO_2. From (8.14(b)) and (8.15) the values of $f(CO_2)$ and $f(CO)$ may be calculated as for the system O–H, giving

$$f(CO_2) = \frac{K P_{gas}\, \gamma(CO)\, \gamma(CO_2)\, f(O_2)^{\frac{1}{2}}}{K \gamma(CO)\, f(O_2)^{\frac{1}{2}} + \gamma(CO_2)} \qquad (8.16)$$

and

$$f(CO) = \frac{P_{gas}\, \gamma(CO)\, \gamma(CO_2)}{K \gamma(CO)\, f(O_2)^{\frac{1}{2}} + \gamma(CO_2)} \qquad (8.17)$$

Log $f(CO)$ as a function of temperature at 1 and 2000 bars is shown in Fig. 8.3 (French and Eugster 1965).

If the $f(O_2)$ of the gas phase in the system C–O is sufficiently low, graphite may precipitate at a 'critical' temperature (see French and Eugster 1965) according to the reaction

$$2CO \rightleftharpoons C + CO_2,$$

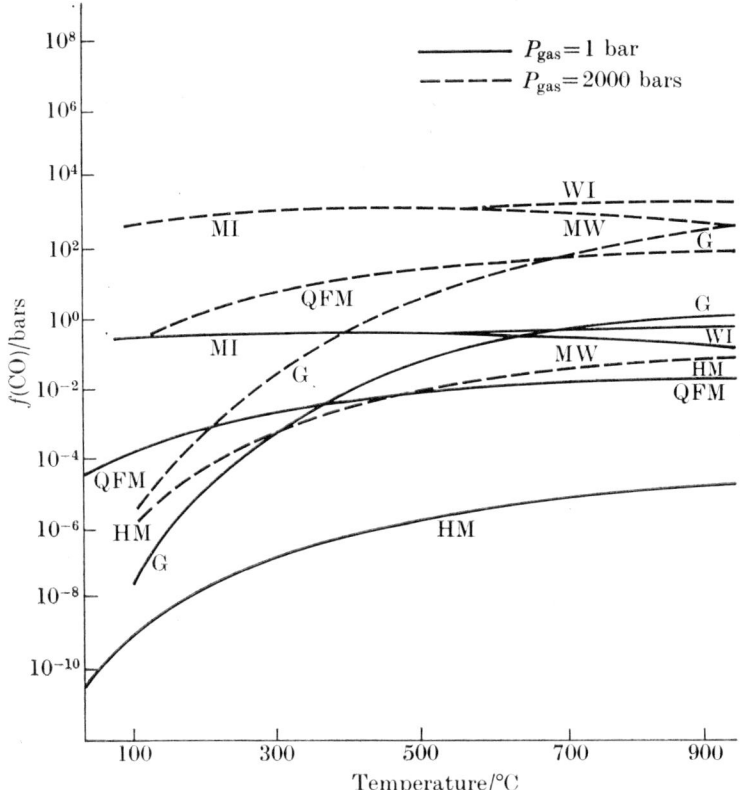

FIG. 8.3. Fugacity of carbon monoxide–temperature plot for graphite and oxygen buffer at 1 bar and 2000 bars total pressure. (After French and Eugster 1965.)

giving equilibrium between the graphite and gas according to the reactions

(a) $C + \tfrac{1}{2}O_2 \rightleftharpoons CO$, where $K_a \dfrac{f(CO)}{f(O_2)^{\frac{1}{2}}}$ (8.18)

(b) $CO + \tfrac{1}{2}O_2 \rightleftharpoons CO_2$, where $K_b = \dfrac{f(CO_2)}{f(CO)f(O_2)^{\frac{1}{2}}}$ (8.19)

The presence of graphite defines the composition of a gas phase and the fugacities of CO_2, CO, and O_2 for a given temperature and total pressure and graphite may therefore be used as an oxygen buffer. From equations (8.14(a)), (8.18), and (8.19) expressions for $p(CO_2)$ and

$p(\mathrm{CO})$ may be obtained (French and Eugster 1965). (These are simplified if $f(\mathrm{O}_2)$ is disregarded as being negligible in comparison to $f(\mathrm{CO}_2)$ and $f(\mathrm{CO})$).

$$p(\mathrm{CO}) = \left[\frac{K_b\,\gamma(\mathrm{CO})^2}{K_a\,\gamma(\mathrm{CO}_2)}\right] p(\mathrm{CO})^2 \qquad (8.20)$$

and

$$p(\mathrm{CO}) = \frac{-1 + \left[1 + 4p(\mathrm{gas})\dfrac{K_b\,\gamma(\mathrm{CO})^2}{K_a\,\gamma(\mathrm{CO}_2)}\right]^{\frac{1}{2}}}{2\dfrac{K_b\,\gamma(\mathrm{CO})^2}{K_a\,\gamma(\mathrm{CO}_2)}} \qquad (8.21)$$

where $\gamma(\mathrm{CO}_2)$ and $\gamma(\mathrm{CO})$ represent the fugacity coefficients of CO_2 and CO respectively. From (8.19), (8.20), and (8.21) the value of $\log f(\mathrm{O}_2)$ for equilibrium with graphite is

$$\log f(\mathrm{O}_2) = 2\left[\log\frac{\gamma(\mathrm{CO}_2)\,p(\mathrm{CO}_2)}{\gamma(\mathrm{CO})\,p(\mathrm{CO})} - \log K_b\right] \qquad (8.22)$$

French and Eugster have calculated $\log f(\mathrm{O}_2)$ for the graphite buffer at various temperatures and total pressures, as shown in Fig. 8.4. Increasing total pressure has a much larger effect on the $f(\mathrm{O}_2)$ values obtained from the graphite buffer than those obtained from oxygen buffers.

Using similar methods, gas fugacities can be controlled in many other two-component gas systems. Eugster and Skippen (1967) describe a method for controlling $f(\mathrm{H}_2)$ in the C–H system using a methane buffer.

A combination of the graphite buffer with various oxygen buffers can be used to control fugacities of gases in the C–O–H system. French (1966) calculated the fugacities of CO_2, CO, $\mathrm{H}_2\mathrm{O}$, H_2, CH_4, and O_2 using this method. If graphite is present in both the charge and buffer systems the fugacities of all gas species are equal (see p. 185 for experimental details). If, however, the components of the gas phase in the buffer and charge systems differ, the $f(\mathrm{O}_2)$ values in the two systems will differ. Eugster and Skippen (1967, pp. 503–6) have determined the composition of a C–O–H gas at 2000 bars total pressure using both methods. Their results are shown in Fig. 8.5(a) and (b), which indicate the radical changes in composition when the components of the gas in the buffer are the same as in the charge (Fig. 8.5(a)) and when the components of the gas in the buffer and charge systems differ (Fig. 8.5(b)).

Partial pressures of volatile components 181

Other buffers. Using the principles just described, a large number of possible solid buffer techniques to control other gas fugacities are possible. Eugster and Skippen (1967) have calculated $f(H_2)$, $f(H_2O)$, $f(H_2S)$, $f(HS)$, $f(S_2)$, $f(SO_2)$, and $f(SO)$ for the assemblages: pyrrhotite + iron + magnetite or wüstite; pyrrhotite + pyrite + magnetite; and hematite + magnetite + pyrite, in the system H–O–S. By adding graphite to this system the fugacities of four-component gases in the system C–O–H–S can be calculated. In such systems, $f(O_2)$, and $f(S_2)$ are

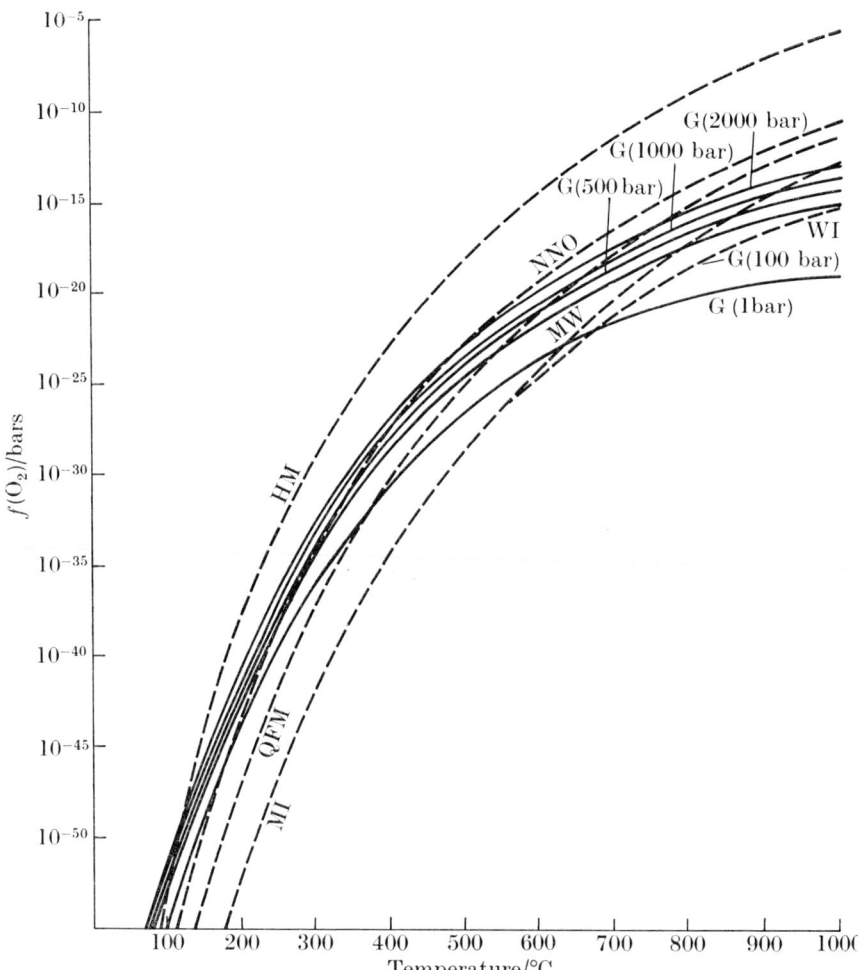

FIG. 8.4. Fugacity of oxygen–temperature plot for graphite buffer at various gas pressures. Graphite buffers shown as solid lines with gas pressures in brackets; dashed curves are values for other buffers. (After French and Eugster 1965.)

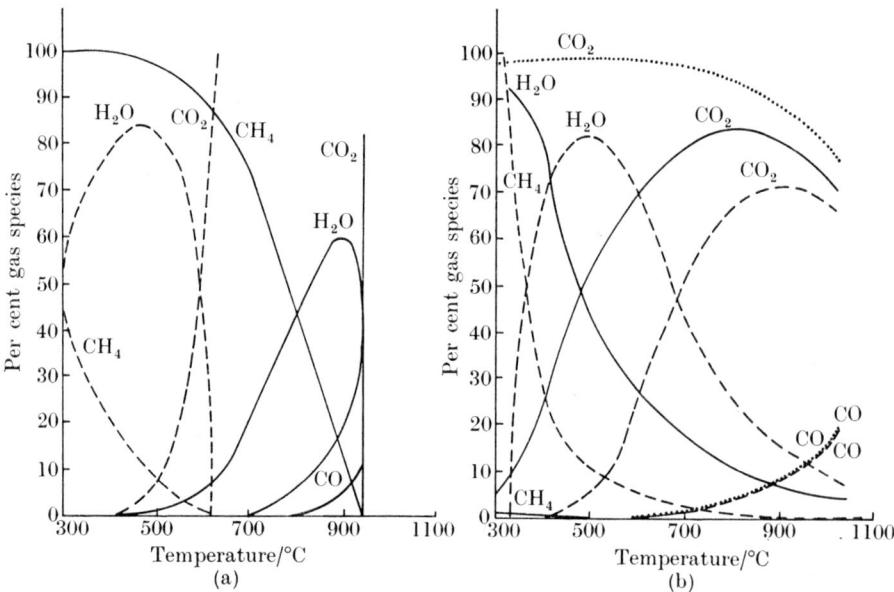

FIG. 8.5. Compositions of gases in the system C–O–H in the presence of graphite at various temperatures and 2000 bars pressure: (a) dashed curve $f(H_2)$ controlled by QFMG, COH buffer; solid curve $f(H_2)$ controlled by MIG, COH and WMG, COH buffers; (b) dotted curve $f(H_2)$ controlled by HM, OH buffer. Solid curve $f(H_2)$ controlled by NNO, OH buffer. Dashed curve $f(H_2)$ controlled by QFM, OH buffer. (After Eugster and Skippen 1967.)

defined internally or $f(H_2)$ is defined externally and $f(S_2)$ internally. Details of these procedures are given in Eugster and Skippen (1967, pp. 511–17).

Munoz and Eugster (1969) have described a method for controlling fugacities in the H–O–F system using fluorine buffer assemblages to control the ratios of HF and H_2O fugacities and $f(H_2)$ externally using a standard oxygen or a methane buffer. The fluorine buffer reactions are:

WFQ
$$CaSiO_3 + 2HF \rightleftharpoons CaF_2 + SiO_2 + H_2O$$
(wollastonite)   (fluorite)   (quartz)

AFSQ
$$CaAl_2Si_2O_8 + 2HF \rightleftharpoons CaF_2 + Al_2SiO_5 + SiO_2 + H_2O$$
(anorthite)   (fluorite)   (sillimanite)   (quartz)

In the presence of graphite, fluorine buffers can be used to calculate gases in the C–O–H–F system according to the reaction

CFG
$$CaCO_3 + 2HF \rightleftharpoons CaF_2 + C + H_2O + O_2.$$
(calcite)   (fluorite)   (graphite)

Partial pressures of volatile components 183

The appropriate equations and details of the fugacity calculations are given by Munoz and Eugster (1969).

Preparations and procedures for buffering techniques

With the exception of a pressure-vessel developed by Shaw (1963) for experiments in which the hydrogen fugacity can be varied continuously (described in a later section), buffered experiments are normally done in standard cold-seal pressure vessels (see Chapter 5), at total pressures generally less than 3 kbar. The principal difference between buffered experiments and those described previously is in the sealed-tube techniques. In general, the greater the number of gas fugacities being controlled, the greater the number of tubes required.

Preparation of buffers. Wherever possible, solid buffer assemblages are prepared from 'spec. pure' chemicals mixed in the proportions required by the equations in this chapter. For some buffers, 'spec. pure' chemicals are unavailable and the buffer components must be prepared before mixing. For example, Huebner and Sato (1970) prepare manganosite by reducing pyrolusite (MnO_2) in a stream of hydrogen at 800 °C. The buffer assemblages must be as pure as possible because the thermodynamic constants from which the gas fugacities are calculated are usually based on pure compounds. Depending on the buffer assemblage being used, about 40–70 mg of solid buffer are required for each experiment.

For the methane buffer (Eugster and Skippen 1967), methane gas is pumped into a pressure-vessel containing a graphite rod.

Capsule preparation. The simplest type of capsule arrangement involves O–H gases using only oxygen buffers. This requires a two- or three-tube technique as illustrated in Fig. 8.6(a). The charge together

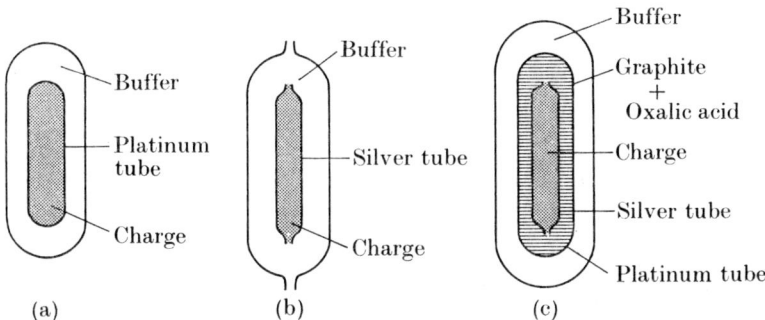

FIG. 8.6. Sample tube technique for buffered experiments: (a) for oxygen buffers in the system O–H; (b) for graphite buffers in the system C–O; (c) for oxygen and graphite buffers in the system C–O–H. (After Eugster and Skippen 1967.)

with water in an approximate 2 : 1 weight ratio is sealed in a platinum capsule about 20 mm long, of 3 mm o.d. and 0·25 mm wall thickness using a d.c. arc welder (see Chapter 5, p. 107). A small unwelded silver tube is placed inside the platinum tube. The platinum tube in turn is placed in a larger gold tube (about 35 mm long, 5 mm o.d., and 0·25 mm wall thickness) containing the oxygen buffer assemblage and 20 to 30 mg of water. The gold tube is then welded.† Tubes of these dimensions will fit into a $\frac{1}{4}$-in i.d. cold-seal pressure-vessel.

The purposes of the various tubes are as follows. The platinum tube, containing the charge and water, separates the bulk composition being investigated from the buffer and water. Platinum is, however, permeable to hydrogen, which can freely pass through the walls of the platinum capsule. Thus, the hydrogen fugacity is fixed in the charge, and the gas fugacities in both the buffer and charge are equal. The platinum capsule also prevents contamination of the charge by the buffer. The small unsealed silver capsule prevents excessive loss of iron by alloying with the platinum capsule. Alternatively, iron loss may be cut down by using a Ag–Pd tube instead of the platinum tube (Muan 1963). Turnock (1960) has used sealed silver tubes instead of platinum tubes. However, the lower permeability of hydrogen through silver, and the lower melting-point of silver (960·8 °C) makes this substitution undesirable. The sealed outer gold tube protects the buffer assemblage from the external pressure (P_{total}) media, which may be H_2O, CO_2, or any other suitable fluid, and from the walls of the pressure-vessel. The external pressure is imposed upon the charge and buffer by collapse of this outer tube.

The capsule arrangement for the graphite buffer, where only gases in the system C–O are present, is shown in Fig. 8.6(b). In this case both the tubes are of silver and are unsealed. The inner tube contains the sample; the outer tube contains the graphite buffer or a suitable oxygen buffer.‡ Whichever buffer is used, the external total pressure is provided by a CO_2, or CO_2–CO gas mixture. Because no suitable metal is known which acts as a semipermeable membrane for these gases, the silver tubes are crimped rather than welded to allow free access of CO_2 and CO to both buffer and charge. The CO_2 (or CO_2–CO mixture) diffuses into the buffer

† The outer and middle tubes used in buffered experiments should be welded by crimping the tube at three points, 120° from one another, along its circumference before welding. This produces a 'cocked-hat' weld by drawing the carbon arc from the outside of the crimp toward the centre. During this procedure the tube is held in a chuck designed for the purpose (see Chapter 5, p. 107).

‡ The oxygen buffer used will depend on the system being investigated. For example an iron-bearing oxygen buffer would be used for a system containing siderite (see French and Eugster, 1965).

Partial pressures of volatile components 185

and the ratio of CO_2 to CO at equilibrium is established by the $f(O_2)$ of the buffer used. In this type of experiment, no water is added either to the buffer or to the charge.

For three-gas systems or systems with more components, involving more than one buffer assemblage, a three-tube technique must be used. A number of variations of this technique are possible. The general method is illustrated in Fig. 8.6(c). For gases in the system C–O–H, the inner system consists of a crimped silver tube containing the charge and a sealed platinum or Pd–Ag tube containing graphite and a suitable source of the C–O–H gas such as oxalic acid, benzoic acid, and water (Eugster and Skippen 1967, p. 507). The outer system contains the buffer sealed in a large welded gold tube. This buffer system may be an oxygen buffer and water or oxygen buffer + graphite + C–O–H gas source.

Using this technique, experiments may be done in which the components of the gas phase are the same in both the charge and buffer systems, or experiments in which the components of the gas phase are different. In the former, the presence of graphite in both charge and buffer allows equilibration of $f(H_2)$ through a membrane, resulting in equal fugacities in both systems. In the latter, a solid oxygen buffer and an O–H gas in the outer (buffer) system imposes a fixed $f(H_2)$ on the inner (charge) system consisting of C–O–H gas in equilibrium with graphite. Thus $f(O_2)$ for the charge system is not the same as for the buffer system. This technique is particularly useful when field boundaries in $f(O_2)$–T space are parallel to buffer curves in the O–H system.

A three-tube technique similar to that shown in Fig. 8.6(c) is used for the fluorine buffer (Munoz and Eugster 1969). In this case the charge is surrounded by fluorine buffer in the inner tubes and the oxygen buffer contained in an outer sealed gold tube and separated from the fluorine buffer charge system by a sealed platinum tube. Unfortunately, fluorine buffers are not yet calibrated and probably cannot be calibrated until $f(F_2)$ can be monitored directly by a method analogous to that for oxygen by a sensing cell as described in Chapter 4.

Because of the large number of possible buffer combinations for controlling fugacities in complex gas systems, Eugster and Skippen (1967, Fig. 1, Table 3) have developed a shorthand notation to show buffer and charge arrangements. This can be illustrated with reference to Fig. 8.6. In Fig. 8.6(a), the simplest case, the charge (X) in equilibrium with O–H gas is surrounded by an oxygen buffer assemblage (O_B) also in equilibrium with O–H gas. This is written as O_B, OH, (X, OH) where the parentheses indicate a sealed-tube. In the C–O system (Fig. 8.6(b)), the charge (X), in equilibrium with C–O gas, is surrounded by either oxygen buffer (O_B) or graphite buffer (G) in equilibrium with C–O gas giving the notation O_B, CO |X, CO| or G, CO |X, CO| respectively; the vertical bars in this case denoting a crimped tube. In the C–O–H system, the charge (X) and graphite (C) are in equilibrium with C–O–H gas, and the oxygen buffer (O_B) separated by a semipermeable membrane is in equilibrium with O–H gas giving O_B, OH (GX, COH).

Run procedure. The charge is placed in a cold-seal pressure vessel at the desired temperature and pressure for a sufficient length of time to attain equilibrium. At the end of the run, the vessel is quenched, the capsules weighed and both the charge and buffer examined by optical and X-ray methods.

Determination of equilibrium in buffer experiments

The general problems associated with attainment of equilibrium in experimental studies are discussed in Chapter 9. In buffered experiments, equilibrium must be attained with respect to both the total pressure and temperature of the experiment and with respect to the gas fugacities between buffer and charge.

Equilibrium between charge and buffer can be demonstrated in a number of ways. Phases present in both buffer and charge, such as magnetite, hematite, etc., indicate equilibrium with respect to fugacity regardless of the initial gas fugacity in the charge. Some phases have properties dependent on the gas fugacity. These can be changed reversibly by equilibrating the phase with different buffer assemblages, thus indicating equilibrium conditions. If platinum is used as a semipermeable hydrogen membrane, runs with sealed and unsealed tubes should be equivalent for a given buffer under the same experimental conditions if equilibrium has been attained. Consistency of data using different buffers is also a good indication of equilibrium.

In all buffer experiments, the buffer should be examined after the run for the disappearance of phases, oxidation, or reduction. Any or all of these will indicate disequilibrium.

The time required for equilibration of gas between buffer and charge depends on the temperature, pressure, and fugacities of the gases involved. High temperatures and pressures produce equilibration in a short time. Ernst (1960) estimates equilibrium of $p(O_2)$ between oxygen buffers and charge takes place in a few hours at 800 °C at a total pressure of 1 to 2 kbar. At low $f(H_2)$ values and low temperatures, e.g. HM buffer at 500 °C and MI buffer at 300 °C, equilibrium is not achieved before the water has oxidized the charge.

Because of oxidation, certain buffers with high $f(H_2)$ values are almost useless. Eugster and Wones (1962, p. 94) show that in the QFI buffer, the quartz and iron tend to be oxidized to fayalite in a few hours using an outer gold tube which is permeable to hydrogen. Reducing the amount of water added to this buffer merely results in the buffer running dry and giving improper $f(O_2)$ values.

In long runs at temperatures above 700 °C the pressure-vessel wall may effect the $f(O_2)$ values owing to reaction between the wall and H_2O. In such cases the vessel wall and the external water providing the pressure act as a buffer owing to the slight permeability of hydrogen through the gold tube. Eugster and Wones (1962) show that a new Stellite 25 vessel has an $f(O_2)$ equal to the fayalite stability region (Fig. 8.1), whereas an old vessel has an $f(O_2)$ slightly greater than the NNO buffer.

Equilibrium of buffer and charge can be demonstrated by doing identical runs at increasing times. Absence of change in the assemblages produced is a good indication that equilibrium has been achieved.

Control of $f(O_2)$ by hydrogen-water vapour mixtures

Shaw (1963, 1967) has described a simple technique, using a modified cold-seal pressure vessel, whereby the fugacities of hydrogen, oxygen, and H_2O can be continuously controlled by diffusion of hydrogen through a suitable membrane. This method, based on the dissociation reaction for water and its resulting equilibrium constant (eqn 8.1), is superior to the oxygen buffer method in that P, T, and $f(O_2)$ can be varied independently and continuously of one another, and $f(H_2O)$ is observed rather than calculated. It is, however, restricted to hydrothermal systems in which H_2O is the only important component in the gas phase in which hydrogen is inert.

The apparatus used is illustrated in Fig. 8.7. In Fig. 8.7(a) the modifications of the cold-seal pressure-vessel are shown. A high-pressure capillary tube, surrounded by the membrane, runs the length of the reaction vessel with the bottom silver-soldered to a lower high-pressure nut which is attached to the hydrogen pressure source. Details of suitable membranes are given by Shaw (1967). At the top end of the assembly the membrane widens and is packed with quartz or fitted with a porous steel plug and welded shut. The part of the membrane which contains the hydrogen is thus of small length and at constant temperature. The sample is placed in a capsule of the same material as the membrane and welded shut. By means of a separate inlet valve, shown in Fig. 8.7(a) and described by Shaw (1967), argon gas is pumped into the remaining volume of the vessel at a higher pressure than the hydrogen gas.

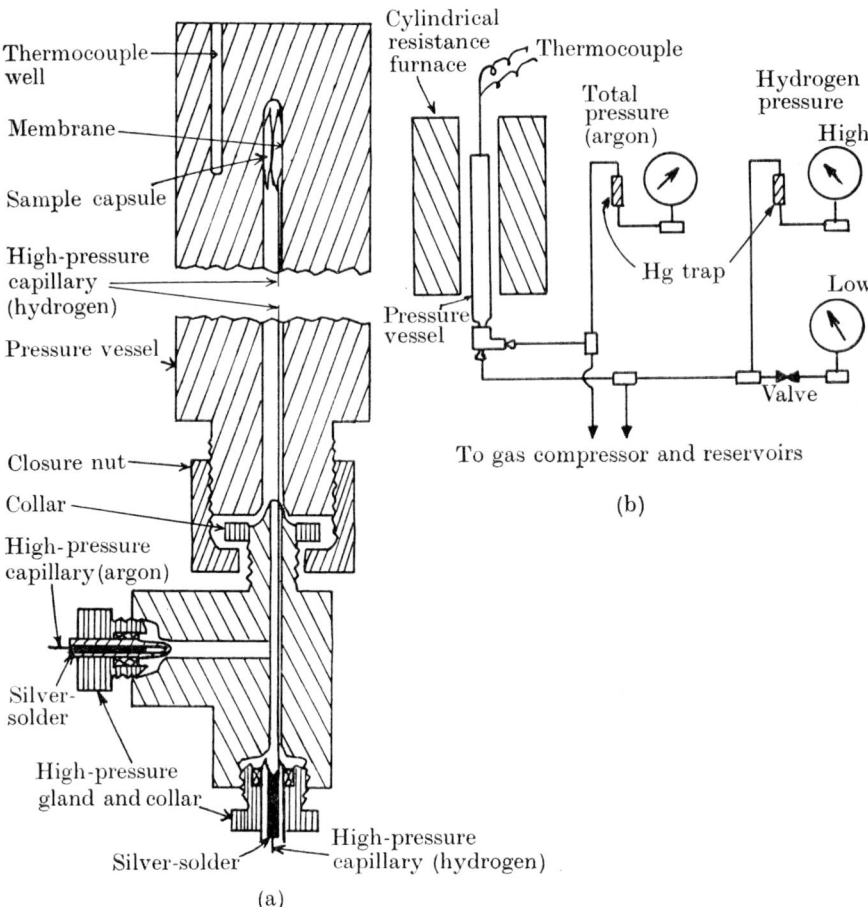

FIG. 8.7. Modified cold-seal pressure-vessel for controlling $f(H_2)$ by a semipermeable membrane: (a) pressure-vessel; (b) pressure and temperature arrangement. (After Shaw 1967.)

If it is assumed that O_2 is a negligible species (see p. 173), the system then consists of two gas species, H_2 (in the capillary and expanded membrane) and H_2O (in the sample capsule), at the same total pressure as the argon gas. Since the argon is at higher pressure than the hydrogen, $p(H_2O) > p(H_2)$ at the beginning of the run. As shown in Fig. 8.7(b) the pressures of both the hydrogen and argon are measured. Details of these measurements are given by Shaw (1967, p. 526–8).

On the assumption that the only gas species occur in the O–H system and any other gas components in the charge and that the rate of loss of hydrogen through the walls of the vessel is negligible, the hydrogen osmotically passes through the membranes until $p(H_2)$ is equal to $p(H_2O)$ and the temperature, total fluid pressure, and hydrogen pressure are measured. In order to calculate $f(H_2), f(O_2)$ and $f(H_2O)$, the fugacity coefficients for H_2 and H_2O and the equilibrium constant for H_2O are required. From this basic data, activities of H_2 and H_2O are determined, a correction is made for the dependence of temperature on the molecular interaction of the gas species, and the value of $f(O_2)$ calculated from

$$f(O_2) = \left[\frac{f(H_2O)}{(f(H_2)K_f}\right]^2. \qquad (8.23)$$

where K_f if the equilibrium constant for water.

An example of such calculation and reference to data on fugacity coefficients and equilibrium constants is given by Shaw (1967).

In experiments of this type it is necessary to establish equilibrium both in terms of the chemical reaction and the rate of diffusion of hydrogen. Reversible equilibrium of the reaction can be demonstrated at a given hydrogen fugacity by using two sample capsules, one containing the reactants, the other the products. Whether osmotic equilibrium between the charge and hydrogen has been achieved may be tested by reversing the flow of hydrogen in different runs from low to high values of a given intermediate pressure, or by changing the direction of flow of hydrogen by introducing a suitable volume of hydrogen into the pure argon. These methods are described by Shaw (1967).

References

DARKEN, L. S. and GURRY, R. W. (1945). *J. Am. chem. Soc.* **67**, 1398.
—— —— (1946). *J. Am. chem. Soc.* **68**, 798.
—— —— (1953). *Physical chemistry of metals*, McGraw-Hill, New York, 1953.
Ernst, W. G. (1960). *Geochim. cosmochim. Acta.* **19**, 10.
—— (1962). *J. Geol.* **70**, 689.
EUGSTER, H. P. (1957). *J. chem. phys.* **26**, 1760.
—— (1959). In *Researches in geochemistry* (ed. P. H. Abelson). Wiley, New York, **1**, 397.
—— and SKIPPEN, G. B. (1967). In *Researches in geochemistry* (ed. P. H. Abelson). Wiley, New York, **2**, 492.
—— and WONES, D. R. (1962). *J. Petrol.* **3**, 82.

FRENCH, B. M. (1966). *Rev. Geophys.* **4**, 223.
—— and EUGSTER, H. P. (1965). *J. geophys. Res.* **70**, 1529.
GREENWOOD, H. J. (1961). *J. geophys. Res.* **66**, 3923.
HOLSER, W. T. (1954). *J. phys. Chem.* **58**, 316.
HUEBNER, J. S. and SATO, M. (1970). *Am. Miner.* **55**, 934.
JANAF tables (1960). *Joint Army, Navy, Air Force tables of thermochemical data*, Dow Chemical Company, Midland, Michigan.
MUAN, A. (1958). *Am. J. Sci.* **256**, 171.
—— (1963). *Am. ceram. Soc. Bull.* **46**, 344.
—— and OSBORN, E. F. (1956). *J. Am. ceram. Soc.* **39**, 121.
MUNOZ, J. L. and EUGSTER, H. P. (1969). *Am. Miner.* **54**, 943.
OSBORN, E. F. (1959). *Am. J. Sci.* **257**, 609.
SHAW, H. R. (1963). *Science* **139**, 1220.
—— (1967). *Researches in geochemistry* (ed. P. H. Abelson). Wiley, New York, **2**, 521.
—— and WONES, D. R. (1963). *Am. J. Sci.* **262**, 918.
TURNOCK, A. C. (1960). *Carnegie Inst. Wash. Year Bk.* **60**, 134.

9. Problems of applying experimental results

Introduction

THE techniques most widely used in experimental petrology have been described in the preceding chapters. In this chapter some problems of interpretation of the experimental results are considered. Because such interpretation depends on the careful identification of the products of each experiment, a brief discussion of methods of identification is included.

The problems related to the interpretation of the results of experimental petrology fall into a number of broad, interrelated categories, the foremost of which is the establishment of equilibrium in the laboratory and in nature. This problem is closely related to kinetics. Another responsibility of the experimentalist is to represent his data in such a way that it can be readily understood by the non-experimentalist. Correlation of the experimental results and those deduced from analyses of the rocks themselves may also be difficult. Here the experimentalist must be constantly aware that his results may represent gross simplifications of natural processes because of the simplified starting-materials, short reaction times, and other limitations discussed in Chapter 1.

Identification of phases

The fine grain-size of most products of experimental studies makes their identification by standard optical techniques difficult in comparison with natural rocks and minerals, because such diagnostic features as colour, extinction angles, and sign of optical rotation cannot be readily determined. Another difficulty is, of course, the limited amount of material available in the standard quenching experiment, which makes the production of thin sections troublesome.

Many of these problems can be overcome by using instrumental techniques, such as X-ray diffraction. However, the petrographic microscope is still the most useful method for identifying experimental products, particularly in liquidus studies and in determining limits of solid solution, where recognition of small amounts of phases is required.

The inability of X-ray diffraction methods to detect glass or crystalline phases present in amounts less than about 5 per cent limits its use in such studies.

Recently, the electron microprobe has been employed to determine the compositions of small amounts of phases where there is no suitable X-ray or optical method.

Optical methods

Optical identification is used for a number of purposes: (a) to identify liquid (as glass) or 'quench' crystals in glass, in a quenched charge, (b) to identify 'expected' crystalline phases and determine, if possible, their compositions, (c) to identify unknown (or 'unexpected') crystalline phases, and (d) to estimate the relative proportions of phases, both crystalline and liquid.

In determination of the liquidus in a synthetic system, optical methods are the only reliable means of identification. In quenching experiments the metal envelope containing the charge is examined for fritting, which indicates a high proportion of glass in the charge. The charge is then carefully removed from the container, care being taken to ensure that all material is taken out. This is particularly important with charges containing a dense crystalline phase, such as magnetite, which tends to sink to the bottom of the charge. Before crushing, a preliminary optical examination may be made under low power. The material is then crushed and mounted on a glass slide in an oil of appropriate refractive index. For charges containing large amounts of glass relative to crystalline phases, an oil with refractive index close to that of the glass should be used in order that even very minor amounts of any crystals present can be easily identified by their (generally) higher refractive index. Similarly, when detection of a small amount of one crystalline phase with different refractive indices is required, the oil used should correspond to that of the second phase. This is the best method of determining limits of solid solution in synthetic systems.

Where observations of textures or other features such as gravitative separation have to be made, a thin section of the charge may be necessary. For samples with little or no glass, one side of the platinum or gold capsule should be peeled off and the exposed material impregnated in Lakeside 70C cement on a microscope slide. This can then be carefully ground to the correct thickness. A similar procedure may be used for glassy charges, although it may be necessary to grind instead of peeling the metal container since the charge tends to stick to the walls.

In many experiments, a knowledge of the composition and ratio of liquid to crystalline phases is required. This is a difficult problem, particularly in experiments where only a small amount of charge is available and its fine grain-size makes modal analysis impractical. In experiments using glass as starting-material, in which the refractive index of each glass is known relative to its bulk composition, the composition and relative amounts of glass and crystals can be obtained provided the composition of the crystals can be determined by optical or other methods (cf. Roedder 1959). Optical determination of the crystalline phase may be made using refractive indices, or occasionally even such properties as interference colours (cf. Osborn and Schairer 1941). In hydrothermal experiments, in which vapour is an additional phase, this method cannot be used to determine compositions since material may preferentially enter the vapour phase. Determination of the proportions and compositions of all crystalline mixtures can only be done optically if there are sufficiently distinctive differences in optical properties of the phases, and variations in optical properties with composition are known.

X-ray diffraction methods

X-ray powder diffraction is used to complement optical identifications and to determine the compositions and relative proportions of crystalline phases. This method is particularly useful for very small crystals. The principal limitation of the X-ray method is that it cannot detect crystals present in amounts smaller than about 5 per cent and is therefore of little use in locating liquidus temperatures or limits of solid solutions. If more than two phases are present this limit of detection decreases.

In most experimental work, the products are too fine-grained for single crystal X-ray studies and therefore for precise crystallographic measurements. Rapid determinations by X-ray methods are normally made using a Debye–Scherrer camera or diffractometer. These techniques are described by Azaroff and Buerger (1958).

Identification is made by comparison of d spacings with those of the ASTM index.† For many minerals, X-ray or combined X-ray and optical methods are available for the determination of the compositions of most silicates and other minerals occurring as solid solutions. These methods are discussed in many mineralogy books (cf. Deer, Howie, and Zussman 1963; Zussman 1967). Compositions cannot, however, be

† This is an index of d spacings and other crystallographic parameters for minerals, inorganic, and organic compounds published by the American Society for Testing Materials. It is continuously revised as new data are available.

determined by X-ray methods if the end-members of the solid solution have very similar unit-cell sizes and shapes.

The relative proportions of different crystals can be determined in a semiquantitative way by X-ray methods. In a two-crystal mixture the amounts of each are approximately proportional to the intensities of their peaks on a diffractogram or of the lines on a powder film. Thus, by comparing the intensities of the peaks from prominent planes, the relative proportions of each phase can be estimated. Care must be taken in the case of crystals representing solid solutions, since in many solid-solution series the intensities of peaks change with compositional changes. Similarly, intensities will change if differential recrystallization and crystallization of certain phases takes place during the experiment.

This, or other suitable methods can also be used to determine whether equilibrium, in terms of the length of the experiment, has been established. In some cases, experiments run under identical conditions but with increased time will show changes in the X-ray diffraction peaks for each experiment, and equilibrium may be assumed to be established when there is no decrease in the intensity of the phase on the diffraction trace with increasing time. In reactions which are extremely slow, for example graphite to diamond at 400 °C, no change will be observed in X-ray intensities irrespective of increased time of the experiment.

High-temperature X-ray cameras and diffractometers can be used to detect solid–solid phase changes including polymorphic transitions. The techniques have been described by Grundy and Brown (1969) and may be valuable in avoiding any ambiguity due to formation of quench phases. These cameras have been used for example, in studies of feldspar transitions (Stewart and von Limbach 1967; Grundy and Brown 1967).

Electron microprobe

The electron microprobe is a valuable tool for identification and chemical analysis of both liquid and crystalline phases, particularly when such phases are too small for optical methods or there is no suitable X-ray method. Its principal value in experimental work is its ability to determine compositions on grains as small as 5 μm. The electron microprobe may also show zoning in crystals not detectable by optical methods. Zoned crystals indicate disequilibrium.

One of the most important applications of the microprobe is its ability to determine and identify glass compositions which can be determined only indirectly by other methods.

For quantitative microprobe analysis suitable standards are required. One further disadvantage in experimental work is the need to prepare polished samples or polished thin sections which must be carbon-coated before analysis. For very fine-grained material, preparation of sample may be difficult.

The general principles of the electron microprobe analysis and techniques of sample preparation are described by Birks (1963).

Miscellaneous problems of identification

In this section a number of miscellaneous problems of identification of synthetic products are considered.

Using the standard quenching technique it is sometimes impossible to cool the sample rapidly enough to prevent crystals forming above their liquidus temperature or vapour phases in the liquid being trapped and crystallizing on cooling. Such crystals are termed 'quench crystals' and must be distinguished from crystals of the same composition which have crystallized below their true liquidus temperature. Substances which have sharp melting-points and form idiomorphic crystals, e.g. diopside, or crystals formed from liquids of low viscosity, are prone to form 'quench crystals'. Optical identification of 'quench crystals' is somewhat subjective but idiomorphic crystals present only in the glass and not as discrete crystals, and fine-grained feathery aggregates of crystals suggest a quench phase.

In systems with immiscible liquids, the liquids normally occur as glassy 'globules', each with a different refractive index. If immiscibility is suspected in the liquid phases, they can be easily detected. The classical example of this is liquids found in silicate–sulphide systems or the immiscible liquids present in high silica compositions of many common rock-forming oxide–silica systems (cf. Greig, 1927).

Location of phase boundaries in synthetic systems using quenching techniques usually calls for optical identification of very small amounts of one phase in large amounts of another phase or phases. With crystalline phases, this is most easily done by distinctive optical properties such as birefringence, relief, etc. For binary systems with limited solid solution (cf. Fig. 2.3(e)), the solidus can most easily be determined by studying initial compositions which will produce approximately equal amounts of both crystalline phases and identifying them by X-ray diffraction methods. For a true binary system this will locate the solidus and eutectic temperature and will thus avoid unnecessary effort in locating the temperature of this boundary curve.

Equilibrium

Although the terms *equilibrium, disequilibrium,* and *stable and metastable equilibrium* are easily defined in theory and their thermodynamic criteria are well known, experimental proof of equilibrium is very difficult. A definition of homogeneous and heterogeneous equilibria has been given in Chapter 2. A system is in stable equilibrium when all the phases involved are stable, and in metastable equilibrium when one or more of the phases involved are metastable.

The importance of assuring equilibrium results in experimental work and its relationship to other factors such as reaction kinetics was briefly discussed in Chapter 1. The establishment of equilibrium or, alternatively a statement that equilibrium was not attained, is probably one of the most important aspects of experimental work since it proves a basis of evaluating the results when applied to natural rocks. Fyfe (1960, p. 565) has summarized this in the statement, 'Where, as has commonly happened, experimental results conflict with inferences based on geological observations, the experimentalist has a special responsibility to scrutinize and state clearly the limitations of his laboratory procedure.'

In this section, methods of determining equilibrium and the relationships between kinetics and starting-materials on equilibrium are considered. Much of the treatment is due to Roedder (1959), who has dealt with equilibrium in dry systems, and Fyfe (1960), who discusses equilibrium in hydrothermal systems. Both writers stress the difficulties in verifying equilibrium in silicate systems. Fyfe and Bischoff (1965) have outlined problems of stability relations of polymorphs in which the differences in free energies are small, making thermodynamic calculations of stability from calorimetric data imprecise. According to these authors the effects of grain-size, nonhydrostatic stress, and impurities may be as important as free energy differences in the stability of polymorphs. Fyfe (1960, p. 557) points out that absolute proof of stability (as opposed to metastability) of an assemblage is impossible. Nevertheless every effort must be made in experimental work to ensure reasonable certainty of equilibrium using the methods and criteria outlined below.

Determination of equilibrium

If two crystalline phases A and B exist under certain pressure and temperature conditions, A may be considered stable relative to B if crystals of A grow while crystals of B disappear. A is then the equilibrium phase. Similarly a boundary curve between A and B can be considered as representing equilibrium conditions only if A can be converted to B and B to A by changes in the pressure or temperature values across the curve. The accuracy of equilibrium determined in this way is related to the gap between the experimentally determined reaction points.

Unfortunately, both these methods involve reaction kinetics and in practice it is impossible to prove equilibrium, since reaction rates (in terms of the normal duration—weeks or even months—of most experiments) are often extremely slow near boundary curves. It is also impossible to prove that the equilibrium is necessarily stable, since there may be some third phase C which has a lower free energy than A or B but does not appear. (The reader will recall that the most stable phase has the lowest free energy; see p. 18). In such cases it is often necessary to rely on geological observation. The various polymorphs of silica are very prone to metastable formation.

Fortunately there are a number of indirect methods which indicate whether a boundary curve depicts equilibrium conditions. In approximate order of importance these are:

(i) Approaching the boundary curve from temperatures or pressures above and below that of the boundary. For example, in a quenching experiment to determine a liquidus, runs should be made using crystals plus glass (low-temperature assemblage) and glass of the same bulk composition (high-temperature assemblage) as starting-materials. If the same results are obtained the reaction is said to be reversible. This is a good indication of equilibrium. However, in the case of reactions involving silicate liquids their well-known tendencies to undercool owing to their high viscosities may make reversibility difficult. Feldspars and other silica-rich compounds show this tendency. In dry systems crystallization of glass may be induced by adding small amounts of some natural, water-bearing glass, such as obsidian, to the sample and sealing both in an evacuated glass tube; the water acts as a catalyst. This technique has been described by Greig and Barth (1938). In hydrothermal systems the problems of sluggish crystallization due to undercooling are often much less severe. Similarly in liquids with low viscosities, generally those with low SiO_2, nucleation of crystals is often rapid.

The presence of water increases crystallization rates and tends to produce larger crystals, probably by a process of recrystallization.† Large crystals grown in this way may be valuable indicators of

† Theoretically, rapid nucleation should produce small crystals. The presence of water presumably promotes solution and precipitation of material, a phenomenon which will be very slow in a dry system.

equilibrium. If only a few of these crystallize at a given temperature, and if they become corroded at a slightly higher temperature owing to their sluggish melting, the equilibrium temperature is 'bracketed'. Corrosion in crystals as an indicator of equilibrium is more fully discussed in the next section. Roedder (1959) suggests that crystals grown in this way should not exceed 20 μm, since larger crystals will melt very slowly above the equilibrium temperature. Even crystals as large as 20 μm will have large surface free energies which will impede establishment of equilibrium.

Determination of equilibrium conditions for phases in which no liquid is present represents a different situation. This is due in part to differences in starting-materials with different free energies (see p. 203) and to the slower reaction rates at lower subliquidus temperatures. In a subliquidus reaction between solids A and B, rates are appreciable only at temperatures and pressures which differ considerably from equilibrium conditions. Reversibility is established by making isobaric (or isothermal) runs to establish the temperature (or pressure) at which A $\rightarrow$ B and B $\rightarrow$ A. (In this case A and B may represent single solid phases, an assemblage of solid phases, or solid and vapour phases, etc.) Using this method the equilibrium conditions are 'bracketed', although for many subliquidus reactions these 'brackets' may be very wide (± 100 °C is not uncommon) because of the sluggishness of reactions close to the equilibrium temperature.

(ii) Increasing the length of the experiment while other variables are kept constant. As discussed above, one of the principal limitations of the experimental method is the inability to duplicate in the laboratory the reaction times available in nature. This is probably the most important cause of metastable equilibrium in synthetic systems. Unfortunately lack of change in the results of experiments, in which all conditions other than time are kept constant, is not a very valuable criterion of equilibrium. In some systems, such as Al_2SiO_5–SiO_2–H_2O containing the important minerals sillimanite, kyanite, andalusite, mullite, pyrophyllite, and clays, reactions are very sluggish and the differences in free energies between phases are very small. In such systems thermodynamic calculations are much more valuable than direct experimental methods. A summary of the problems of establishing equilibrium in the Al_2SiO_5–SiO_2–H_2O system is given by Fyfe and MacKenzie (1969). Similar problems arise in determining the stability of calcite and aragonite, as discussed by Fyfe and Bischoff (1965) and Bischoff and Fyfe (1968).

Another method of indicating whether reaction times are sufficient for equilibrium is to use a different type of charge or one crystallized in a different manner, but keeping the other conditions constant. If a different type of starting-material is used, the relative free energies of the different starting-materials and the conditions of the experiment must be considered. Depending on whether liquid–solid or solid–solid equilibrium is involved, certain types of starting-materials may actually increase the likelihood of metastability. This is discussed in a later part of this chapter.

Reaction rates tend to be particularly sluggish near boundary curves. When a liquid or vapour is involved, intermediate grinding and reheating of a charge may increase the reaction rates by increasing the surface area of the solid available for reaction by the liquid or vapour phase. Theoretically the free energy G of a substance is related to its grain-size by the equation

$$\left(\frac{\partial G}{\partial A}\right)_{PT} = \sigma, \tag{9.1}$$

where A is the surface area and σ the surface tension. Equation (9.1) may be expanded as

$$G_r = G^o + 2\sigma V/r \tag{9.2}$$

where G_r is the free energy of small spherical grains of radius r, G^o the free energy of macrocrystalline material, and V the molar volume.

As pointed out by Fyfe and Bischoff (1965, pp. 5–6) little experimental data are available which relate free energy to grain-size. However, they estimate, for the calcite–aragonite transition, that the last term in eqn (9.2) alters the free energy significantly only within small ranges of r. They emphasize that classical thermodynamics cannot predict the behaviour of very small groups of atoms and that care must be taken when using standard state or equilibrium data (which refers only to material in which surface-energy contributions are insignificant) in considering relative stabilities of polymorphs with fine grain-sizes.

(iii) Textural evidence is a further useful but rather unreliable criterion of equilibrium. The presence of small idiomorphic crystals evenly distributed throughout the quenched charge may indicate equilibrium, whereas rounded or embayed crystals indicate that the duration of the run is inadequate (Schairer 1959). Similarly, zoning or any other obvious inhomogeneity in a crystalline phase indicates disequilibrium. Presence of gas bubbles or liquid entirely enclosed in a solid phase suggest disequilibrium, since the gas or liquid within the crystal is presumably not in equilibrium with the liquid and gas outside the solid phase. However, the presence of blebs of one liquid in a host liquid of a different composition indicates liquid immiscibility, which may represent equilibrium or disequilibrium between liquid phases of different compositions.

(iv) Finally, conformity of data may indicate equilibrium. For example discrepancies in experimental results between closely related compositions or in results from the same composition are a strong indication that disequilibrium is present or that the starting-material is not of the correct composition.

Although these indicators of equilibrium have been discussed principally for synthetic systems, the same criteria should be applied to experiments using natural starting-materials.

Equilibrium and kinetics

It should be clear from the foregoing that failure to attain equilibrium is mainly due to kinetic factors. Fyfe (1960) and Fyfe and Bischoff (1965) have outlined some of the factors involved in synthesis experiments and their relationship to reaction kinetics and equilibrium.

It can be shown thermodynamically (Chapter 2) that the stable phase(s) in a reaction must be the one(s) with the lowest free energy. In a synthesis experiment the starting-materials must therefore have a higher free energy than the phase(s) to be synthesized, if the synthetic products are to exist in their equilibrium state. Thus, the net result must be a decrease in the free energy of the system. These factors are illustrated in Fig. 9.1, where free energy is plotted against temperature. The upper curve represents the free energy of the starting-material, the lower curves the free energies for some typical dehydration reaction:

$$A.H_2O \rightleftharpoons B + H_2O$$

The temperature at which the lower curves cross represents the equilibrium temperature where the free energy of products–reactants (or ΔG) is zero (see Chapter 2, p. 18). Fig. 9.1 may therefore be considered as a type of equilibrium phase diagram.

In any reaction the rate process depends chiefly on the following factors: (a) rate of nucleation; (b) growth rate of the product; (c) solution rate of the reactant and rate of transfer of material by diffusion; (d) recrystallization rate of fine-grained or metastable products. The initial phase which forms in an experiment depends on the relative magnitude of factors (a) and (b) and not on the equilibrium temperature. These factors also determine whether a metastable or stable product will form. If the free energies of the reactants and products have similar values, corresponding to curves in Fig. 9.1 with similar slopes, then the chances of metastability are much greater. In order that a phase may form there must be sufficient supersaturation (or nucleation) of that phase before the rate of growth can play a significant part. For many reactions nucleation and growth are the slowest steps. Considering the lower curves in Fig. 9.1. again, the rate of nucleation, and therefore of growth, of $A.H_2O$ relative to $B + H_2O$ may be correlated approximately

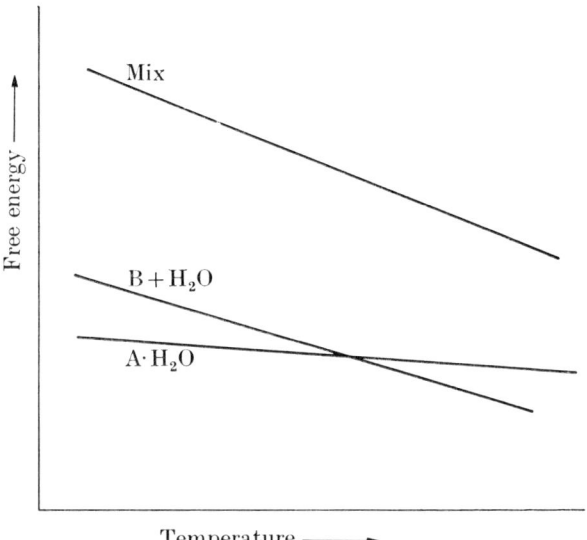

FIG. 9.1. Free energy–temperature plot for starting-materials and a hypothetical reaction. (After Fyfe 1960.)

with the difference in their free energies. This explains why transformations near the equilibrium temperature, where the free energy curves coincide, tend to be very sluggish (see p. 197), whereas further away from the equilibrium temperature the transformation is much more rapid and the chances of producing a stable phase are much better.

The difficulties of obtaining by synthesis a true equilibrium boundary in a transformation follows from Ostwald's law of stages, which indicates that phases tend to appear in the reverse order of their true stability.

In experiments carried out in the presence of water, or some other catalytic agent, the rate of solution of the reactant and transfer of material by diffusion may, in some cases, determine the overall reaction rate. Water acts as a catalyst by promoting solution, transfer of material, and formation of complex ions intermediate between those of reactant and product. For example, in experiments in which silica is one of the reactants it may be much more soluble than the other reactants and can be transferred in solution to new sites of nucleation and growth. Water may also promote the formation of complex ions, but increase the probability of formation of the desired compounds. The solubility of various silicates in supercritical water and transfer of ions in silicate liquids have been studied by many workers (e.g. Anderson and Burnham 1965; Adams 1968; Currie 1968; Orville 1963; Morey and Hesselgesser 1951).

Although the presence of an aqueous phase greatly increases rates of transfer of material, diffusion can also take place in the absence of an aqueous or other catalytic agent. In such cases diffusion may be along grain boundaries or through crystal lattices. In the latter case the rate of transfer by diffusion is generally more rapid in one crystallographic direction than in other directions. For some transformations of this type (see Fyfe and Bischoff 1965, pp. 9–10), the presence of crystal imperfections may decrease already very slow reaction rates.

The amount of supersaturation required to nucleate reaction products is an indication of the differences to be expected between the equilibrium temperature and the temperature at which the reaction will run at a measurable rate. This temperature difference is also inversely proportional to the entropy of the reaction.

Increasing water vapour pressure often appears to increase reaction rates for a constant temperature. This may be due to increase in the density of the water vapour phase causing increased solubility of reactants, and to changes in the dielectric properties of water with increasing density (Fyfe, Turner, and Verhoogen 1958).

The fourth factor in reaction kinetics, recrystallization of fine or metastable products, has the effect of providing new surfaces for reaction to occur and destroying any surficial coatings which may prevent the formation of the stable product. Grinding in laboratory experiments considerably increases reaction rates by increasing surface areas, by providing dislocations which promote growth and, in the case of experiments in the presence of water, by promoting diffusion of material along grain surfaces rather than diffusion through crystal lattices. This factor has been discussed above.

Although of more importance in experiments with natural material than with synthetic starting-materials, the presence of impurities in both the solid and aqueous solutions used may affect rates of nucleation and hence overall reaction rates. Bischoff and Fyfe (1968) have demonstrated that CO_2 and $CaCl_2$ catalyse the aragonite–calcite transformation whereas KOH, Mg^{2+}, and SO_4^{2-} inhibit the crystallization of calcite. Bischoff (1968) has done similar experiments on the crystallization of aragonite during the vaterite–aragonite transformation.

The overall rate of a reaction is determined by the slowest of the four main processes mentioned above. Unfortunately, knowledge of the mechanisms of most geological reactions and hence the kinetic processes involved are very poorly understood. Fyfe et al. (1958), Fyfe (1960), and Fyfe and Bischoff (1965) have given an excellent account of the theory of reaction kinetics and have described a number of examples of petrological importance.

Equilibrium and starting-materials

The choice of starting-materials is also a very important factor which determines whether metastable or stable equilibrium is attained. In the previous section it was noted that large differences in the free energies of reactants and products tended to speed up reactions, whereas small differences made reactions sluggish. The time available in the laboratory for a geological reaction is inevitably many orders of magnitude less than the time available in nature. There is therefore a tendency to use highly reactive starting-materials with much larger free energies than the phases to be synthesized. Unfortunately, such reactive materials are much more likely to produce metastable products than starting-materials with free energy values closer to those of the desired products.

Coprecipitated gels, oxide mixes, and glasses are three widely used starting-materials (Chapter 3). Of these, gels are the most reactive but also tend to synthesize phases metastably at temperatures well below their known stability fields owing to the ease of nucleation of metastable phases in gels, which is a much faster process than the transformation of the metastable to the stable phase. In many cases, investigators have prepared gels which are seeded or dried and are therefore most likely to contain nuclei even before the experiment. Fyfe (1960) gives a number of examples in which starting-materials, as gels or oxide mixes, of a single system give widely differing results indicating the importance of starting-materials on the products of synthesis.

In contrast, glasses have a much lower free energy and being less reactive do not tend to produce metastable phases, particularly near their melting points, although their reactions are generally much slower, particularly at subsolidus temperatures. Glasses are therefore much more suitable than gels for high-temperature liquidus experiments in which there is less likelihood of metastability and the increased temperature promotes faster reaction. In the absence of liquid the choice of starting-material is much more difficult. Highly reactive gels or oxide mixes may produce metastable products, whereas glasses may not react in a reasonable length of time.

Fyfe (1960) suggests a number of ways of partially eliminating this difficulty by using as many different starting-materials of the same bulk composition as possible, and by avoiding starting-materials with very large free energies relative to products (or very fine-grained materials with high surface energies). In order to avoid the high free energy problem, seeds of the assumed stable phase or phases may be added to the starting-material. However, using this procedure care must be taken in interpreting the results of experiments of short duration, for the 'seeded' phases will nucleate and grow at different rates. The problem of metastability using reactive starting-materials may also be reduced by approaching equilibrium conditions from material 'seasoned' by holding it above and below the expected equilibria conditions. It must be stressed that these suggested procedures should be used with caution and that in any mixture of macrocrystalline material which fails to react across a boundary, say

$$A + B + C \rightleftharpoons D + E + F,$$

all experiments done with the particular type of starting-material or the particular composition must be treated with suspicion.

Other methods of achieving equilibrium

The problems of determining equilibrium have been described only for the more commonly used experimental methods. In the past few years new techniques have been developed including high-pressure synthesis, vapour pressure measurements, differential thermal analysis, solubility measurements, and calorimetry.

For simple hydrothermal systems it is sometimes possible to follow a transformation by direct measurement of vapour pressure. Although there are many difficulties associated with this technique, the early work of Giauque (1949) and others has shown the influence of grain size on equilibrium boundaries and that equilibrium temperatures are lower than those obtained by direct synthesis.

One method of avoiding the problem of high free energies of starting-materials relative to products is to use mixtures of minerals as starting-materials and determine the conditions under which certain minerals remain and others are destroyed. In theory in a mixture of A and B there is no nucleation of either phase and the time required for reaction is associated solely with the transfer of material. In practice, most reactions involving mixtures of natural minerals are extremely sluggish and the equilibrium temperature can be approximated only by determining the temperatures at which one phase or assemblage of phases grows while another phase or assemblage of phases disappears; the latter gives a maximum temperature limit of stability.

The problem of slow reaction rates using natural material can be alleviated if an accurate method of determining the relative solubilities of phases is available which provides an indication of the direction in which a reaction is going. Such methods are based on the fact that the most stable phase in a reaction has the lowest solubility. Three techniques which have been used to measure solubilities are solution-conductance (Jamieson 1953) and solution-equilibration and nucleation (Bischoff and Fyfe 1968) for studying calcite–aragonite equilibria; and a single-crystal technique (Evans 1965) for studying the breakdown of muscovite, and muscovite and quartz.

In rare cases, differential thermal analysis can be used to determine equilibrium curves provided the reactions are very rapid as, for example, the α–β quartz. The more sluggish the reaction, the greater the error in determining equilibrium temperatures. A closely analogous method to D.T.A. is to determine transitions using an X-ray diffractometer and heating stage. This technique is described by Grundy and Brown (1969).

Calorimetry may also be used to determine equilibrium conditions. This technique involves experimentally determining heats of reaction, heat capacities of reactants and products (providing entropy data), molar volume of reactants and products and, in the case of a gas phase, knowledge of P–V–T relations. Using these, the value of ΔG may be obtained from eqn 2.7, i.e. $\Delta G = \Delta H - T\Delta S$. Unfortunately, the heats of reaction for many silicates have large uncertainties in part due to experimental difficulties and in part to difficulties in obtaining samples of sufficient purity (producing large discrepancies between samples of the same material). This results in uncertainties in ΔG values of similar magnitude to the actual values. For non-silicate reactions of geological interest the situation is less serious and calorimetric data can be used to calculate which phase(s) is stable. Fyfe and Bischoff (1965) outline the calculations for relative stabilities of calcite and aragonite. For reactions involving gases, which have large entropies at low pressures, errors in thermal data are minimal and calorimetric methods can be superior to direct synthesis in determining equilibrium.

Until recently, calorimetric and derived thermodynamic data were scarce. However, accurate data for silicates and other compounds of gelogical importance are now available (Robie and Waldbaum 1968).

Correlation of experimental results with natural data

The responsibility of the experimentalist to realize and state clearly the limitations of his experiments has been stressed. One of the most important of these is to correlate laboratory results with field observations. When making such correlations one must remember that the field geologist may have only limited knowledge, if any, of experimental techniques and the interpretation of the results should not be 'stretched'.

The results of experimental work are usually presented as phase diagrams, with simple mineral constituents as their components, or as reactions representing what is believed to take place in nature. Because laboratory systems are simplified versions of natural processes, correlation of laboratory results with natural data may be difficult. Terms such as 'the granite system' or 'basalt system' mean systems which approach the composition of granites and basalts, respectively, a point which must be clearly understood by the non-experimental petrologist. For example, 'the granite system' is the system quartz–albite–orthoclase $\pm$ H_2O, which contains no anorthite, mica, or amphibole commonly found as minor but nevertheless important constituents of granitic rocks.

For igneous rocks one common method of correlating experimental results with synthetic systems is to use normative or modal compositions, and to compare the density distribution of such compositions with conditions of univariancy or invariancy of the synthetic system (cf. Tuttle and Bowen 1958; Hamilton and MacKenzie 1965). If there is close correspondence between normative compositions and, say, a minimum or eutectic on a phase diagram then certain inferences regarding the genesis of the rock may be made. However, since the normative or modal constituents of almost all igneous rocks are greater than the number of mineral components of a synthetic system, a decision must be made about the minimum percentage of the normative or moda constituents which is represented by the synthetic systems, and which must also be present in the rock in order to justify using the system to explain the genesis of the rock. If H_2O is a component of a synthetic system under pressure, it must also be clearly stated whether the system represents $p(H_2O) = P_{total}$ or $p(H_2O) < P_{total}$. This is particularly important if melting temperatures are being inferred from synthetic systems since it has been suggested that many granitic rocks (for example) are not formed from water-saturated melts and therefore exist under conditions of $p(H_2O) < P_{total}$, (Fyfe 1970; Brown and Fyfe 1970).

For most igneous rocks chemical analyses are more abundant than modal analyses, and hence normative constituents are most frequently used where the genesis of a rock group (as opposed to a suite of rocks or a single occurrence) is being made. Following the practice of Tuttle and Bowen (1958), a minimum of 80 per cent or more of the mineral components in the synthetic system should occur as normative minerals in the rock if the synthetic data are used to explain the genesis of the rock. Great care should be taken when one of the normative minerals contains an oxide not represented by the synthetic system, e.g. including normative acmite when only diopside is present in the synthetic system (Edgar 1964; Nolan 1966). Because the crystal $\rightleftharpoons$ liquid processes in synthetic systems represent only a portion of the same processes inherent in the normative calculation (CIPW), caution must be taken to ensure that the normative processes are the same as those used in the synthetic ones. For examples, the reader is referred to Carmichael and MacKenzie (1963) and Bailey and Schairer (1966).

Care must also be taken in contouring the distribution of normative plots on the synthetic system. Methods of contouring are described by Chayes (1950).

The problem of presenting results of metamorphic reactions are just as severe. In nature most metamorphic reactions are complex and may involve coupled reactions (see Fyfe et al. 1958). Such reactions may not be directly observed in the laboratory, i.e. A may react to form B without any evidence of an intermediate phase C. However, if the experimentalist suspects that an intermediate phase may be involved he should clearly state it in presenting his results.

Like the igneous minerals, metamorphic minerals are often complex solid solutions, much simplified in the laboratory experiments, e.g. chlorites, micas, amphiboles. The effects of components not represented in the laboratory reactions should, wherever possible, be estimated.

In many aspects laboratory investigation of metamorphic reactions are much more complex than igneous reactions because of the slow kinetics; the presence of stress minerals in metamorphism, which are hard to duplicate in the laboratory; and frequently the importance of more than one volatile in metamorphism. Estimates of these variables should also be considered.

In the past decade many experiments pertinent to both igneous and metamorphic rocks have been carried out with natural rocks. While these have the advantages of eliminating the problems caused by the simplicity of synthetic systems, the interpretation of the results of such experiments must be made with caution. Most rock compositions contain more than five components and therefore the complete phase relations cannot be represented by graphical methods but must be depicted as a series of flow diagrams (Chapter 2). If comparison is to be made with synthetic systems care must be taken both in choosing the appropriate components of the rocks, and in the interpretation of the results of melting and crystallization of the rock when compared to the results of the synthetic system. This is particularly important if estimates of the temperature of crystallization or melting of the rock are being made by comparing observed temperatures with those obtained from phase diagrams. One important consideration here is that pressure of natural rock formation can usualy only be estimated. Comparison of temperatures obtained from rock-melting experiments are often in good agremeent with those of synthetic systems and from natural geothermometers, e.g. feldspars.

References

ADAMS, J. B. (1968). *Am. Miner.* **53**, 1603.
ANDERSON, G. M. and BURNHAM, C. W. (1965). *Am. J. Sci.* **263**, 494.
AZAROFF, L. V. and BUERGER, M. J. (1958). *The powder method*, McGraw-Hill, New York, p. 342.
BAILEY, D. K. and SCHAIRER, J. F. (1964). *Am. J. Sci.* **262**, 1198.
BIRKS, L. S. (1963). *Electron probe microanalysis*, Interscience, New York, p. 253.
BISCHOFF, J. L. (1968). *Am. J. Sci.* **266**, 80.
—— and FYFE, W. S. (1968). *Am. J. Sci.* **266**, 65.
BROWN, G. C. and FYFE, W. S. (1970). *Contr. Miner. Petrol.* **28**, 310.
CARMICHAEL, I. S. E. and MACKENZIE, W. S. (1963). *Am. J. Sci.* **261**, 382.
CHAYES, F. (1950). *Trans. N.Y. Acad. Sci.* Ser. 2, **12**, 141.
CURRIE, K. L. (1968). *Am. J. Sci.* **266**, 321.
DEER, W. A., HOWIE, R. A. and ZUSMAN, J. (1963). *Rock-forming minerals*. (5 vols) Longmans, London.
EDGAR, A. D. (1964). *Am. Mineral.* **49**, 573.
EVANS, B. W. (1965). *Am. J. Sci.* **263**, 647.
FYFE, W. S. (1960). *J. Geol.* **68**, 553.
—— (1970). *Geol. J. (Spec. Issue)*, **2**, 201.
—— and BISCHOFF, J. L. (1965). *Soc. of Econ. Paleontologists and Mineral.*, Spec. Publ. 13, 3.
—— and MACKENZIE, W. S. (1969). *Earth Sci. Rev.* **5**, 185.
—— TURNER, F. J. and VERHOOGEN, J. (1958). *Geol. Soc. Amer., Mem.* 73, p. 259.
GIAUQUE, W. F. (1949). *J. Am. chem. Soc.* **71**, 3192.
GREIG, J. W. (1927). *Am. J. Sci.* **13**, 1.
—— and BARTH, T. F. W. (1938). *Am. J. Sci.* A **35**, 93.

GRUNDY, H. D. and BROWN, W. L. (1967). *Schweiz. Mineral. Petrol. Mitt.* **47**, 21.
—— —— (1969). *Mineral. Mag.* **37**, 156.
HAMILTON, D. L. and MACKENZIE, W. S. (1965). *Mineral. Mag.* **34**, 214.
JAMIESON, J. G. (1953). *J. chem. Phys.* **21**, 1385.
MOREY, G. W. and HESSELGESSER, J. M. (1951). *Econ. Geol.* **46**, 821.
NOLAN, J. (1966). *Quart. J. Geol. Soc.* **122**, 119.
ORVILLE, P. M. (1963). *Am. J. Sci.* **261**, 201.
OSBORN, E. F. and SCHAIRER, J. F. (1941). *Am. J. Sci.* **239**, 715.
ROBIE, R. A. and WALDBAUM, D. R. (1968). *U.S. Geol. Surv. Bull.* 1259, p. 256.
ROEDDER, E. (1959). *Physics and chemistry of the earth*, Pergamon Press, London, **3**, 224.
SCHAIRER, J. F. (1959). In *Physicochemical measurements at high temperatures*. Butterworths, London, 117.
STEWART, D. B. and VON LIMBACH, D. (1967). *Am. Mineral.* **52**, 389.
TUTTLE, O. F. and BOWEN, N. L. (1958). *Geol. Soc. Amer., Mem.* 74, p. 153.
ZUSSMAN, J. (1967). *Physical methods in determinative mineralogy*. Academic Press, London, p. 514.

Appendix: units and dimensions

(a) The physical units in the text and figures of this book are usually those quoted by the original authors. The following table allows conversion of these units. In most current scientific work SI units are used.

	British	*SI*
Pressure	1 (lbf in^{-2})	6·895 Pa(Nm^{-2})
	1 psi	0·07031 kgf cm^{-2}
	1 atm	101 325 Pa
	1 atm	1·0332 kgf cm^{-2}
Temperature	1 °F	(5/9)K
Length	1 in	25·4 mm
	1 ft	0·3048 m
Area	1 in^2	6·4516 × 10^{-4} m^2
Mass	1 lb	0·4536 kg

(b) Wire diameters are frequently quoted in B and S units. The following permits conversion from one to the other.

B and S gauge	*Diameter (mm)*	*B and S gauge*	*Diameter (mm)*
10	2·588	26	0·404
12	2·052	28	0·320
14	1·628	30	0·254
16	1·290	32	0·203
18	1·024	34	0·160
20	0·813	36	0·127
22	0·643	38	0·102
24	0·510	40	0·079

Author Index

Italicized numbers indicate reference pages

ABBOT, L. H., 122, *150*
ADAMS, F. D., 7, 8, *11*
ADAMS, J. B., 202, *208*
ADAMS, L. H., 6, 7, *11*, *12*, 119, 124, 126, *149*, *150*, 164, *168*
ALTHAUS, E., 100, 112, 113, *117*
ANDERSON, G. M., 202, *208*
AZAROFF, L. V., 193, *208*

BAILEY, D. K., 33, *34*, 207, *208*
BARNES, H. L., 10, 115, *117*
BARRER, R. M., 52, *65*
BARTH, T. F. W., 13, 22, 29, *34*, 197, *208*
BELL, P. M., 151, 162, 163, 164, 165, 166, *168*, *169*
BIGGAR, G. M., 52, 55, 56, 58, 59, 60, 64, 65, *65*, 76, 82, 88, *93*, *93*
BISCHOFF, J. L., 196, 198, 199, 200, 202, 203, 205, *208*
BIRCH, F., 122, 124, *149*, 164, *168*
BIRKS, L. S., 195, *208*
BLANDER, M., 34, *34*
BOETTCHER, A. L., 162, 163, 164, *168*, *169*
BONNICHSEN, J. R., 164, *169*
BOWEN, N. L., 6, 7, 8, *11*, *12*, 22, 29, *34*, 41, 44, 45, 48, 49, 65, 66, 206, 207, *209*
BOYD, F. R., 9, *12*, 151, 155, 156, 157, 162, 164, 165, *168*
BRADLEY, C. C., 151, 159, 160, 161, *168*
BRADLEY, R. S., 124, *149*
BRIDGMAN, P. W., 8, 9, *12*, 95, 124, 126, 129, 130, 134, 135, 139, *149*, 151, 152, 155, *168*
BROWN, G. C., 207, *208*
BROWN, W. L., 194, 205, *209*
BUERGER, M. J., 193, *208*
BURNHAM, C. W., 108, *118*, 119, 120, 122, 133, 134, 136, *149*, 202, *208*

CARMICHAEL, I. S. E., 207, *208*
CERNYCH, V. V., 8, *12*
CHAYES, F., 207, *208*
CHIPMAN, D. W., 163, *169*
CLARK, S. P., *34*, 122, *149*
CLARKE, F. W., 5, *12*
COES, L., 9, 155, *169*

CROSS, W., 5, *12*
CURRIE, K. L., 115, *118*, 202, *208*

DACHILLE, F., 152, 153, 154, 155, *169*
DARKEN, L. S., 88, 89, 92, *93*, 170, *189*
DAVIS, N. F., 119, 133, 134, *149*, *150*
DAVIES, I., *118*
DEER, W. A., 22, *34*, 193, *208*

EDGAR, A. D., 207, *208*
EHLERS, E. G., 61, *65*
ENGLAND, J. L., 9, *12*, 151, 155, 156, 157, 162, 164, 165, *168*, *169*
ERNST, W. G., 55, *65*, 174, 175, 176, 186, *189*
ESKOLA, P., 40, *65*
EUGSTER, H. P., 9, 10, 171, 172, 173, 174, 175, 177, 178, 179, 180, 181, 182, 183, 184, 185, 187, *189*, *190*
EVANS, B. W., 205, *208*
EYLES, V. A., 4, *12*

FAUST, G. T., 79, *93*
FAWCETT, J. J., 99, 101, 112, *118*
FENNER, C. N., 6, 7, *12*, 94, 95, *118*
FINDLAY, A., 13, *34*
FORSYTHE, W. E., 76, *93*
FRANCO, R. R., 7, *12*
FRENCH, B. M., 172, 178, 179, 180, 181, 184, *190*
FRONDEL, E., 40, 55, *65*, 66
FYFE, W. S., 3, *12*, 13, *34*, 37, 38, 39, *65*, 196, 198, 199, 200, 201, 202, 203, 204, 205, 207, *208*

GANGULI, D., 40, *65*
GETTING, I. C., 162, 165, 166, *169*
GIAUQUE, W. F., 204, *208*
GILBERT, M. G., 162, *168*
GLASSTONE, S., 13, *34*, 171, *195*
GOLDSMITH, J. R., 48, *65*, 119, 120, 126, 127, 128, 129, 142, 143, *150*
GORANSON, R. W., 6, 7, *12*, 97, *118*, 119, 147, *150*
GORDON, T. M., 96, 97, *118*
GREEN, D. H., 162, *169*
GREENWOOD, H. J., 96, 97, *118*, 171, *190*

GREIG, J. W., 6, 7, 12, 40, *65*, 197, *208*
GRIGGS, D. T., 126, 152, *169*
GRUNDY, H. D., 194, 205, *209*
GURRY, R. W., 88, 89, 92, *93*, 170, *189*

HADIDIACOS, C. G., 106, *118*
HALL, H. T., 9, 155, 159, 161, *169*
HALL, J., 3, 4, *12*
HAMILTON, D. L., 13, 40, 52, 53, 55, 56, 60, *66*, 206, *209*
HARRIS, P. G., 112, 113, 114, 116, *118*
HAYS, J. F., 163, *169*
HEALD, E. F., 55, 57, *66*
HEARD, H. C., 119, 120, 126, 127, 128, 129, 142, 143, *150*
HENDERSON, C. M. B., 40, 52, 53, 55, 56, 60, *66*
HERRINGTON, D. R., 55, *66*
HESSELGESSER, J. M., 96, 117, *118*, 202, *209*
HIGNETT, T. P., 48, *66*
HILLEBRAND, W. F., 5, *12*
HLABSE, T., 164, *169*
HOLLOWAY, J. R., 97, *118*, 119, 120, 122, 124, 133, 134, 135, 136, 139, 141, 143, 145, *150*
HOLSER, W. T., 173, *190*
HOWIE, R. A., 22, *34*, 193, *208*
HUEBNER, J. S., 172, 174, 183, *190*
HUTTON, V. O., 161, *169*

IDDINGS, J. P., 5, *12*
INGAMELLS, C. O., 52, 53, 54, 55, 60, *66*
INGERSON, E., 40, *66*, 94, *118*
ITO, J., 40, 55, *65*, *66*

JAHNS, R. H., 108, *118*
JAMES, R. S., *118*
JAMIESON, J. G., 205, *209*
J.A.N.A.F., 173, *190*
JEFFERY, R. N., 146, *190*
JOHANNES, W., 162, 163, 165, *169*
JOHNSON, D. P., 128, *150*, 161, *169*
JOHNSTON, J., 124, *149*
JURA, G. H., 154, *169*
JURA, O., 154, *169*

KEENE, A. G., 48, *66*
KELLEY, K. K., 164, *169*
KENNEDY, G. C., 9, *12*, 96, 118, 152, 162, 164, 165, 166, *169*
KERN, R., 13, *34*
KIEFFER, J., 151, *169*
KING, E. G., 164, *169*
KITAHARA, S., 162, *169*
KLEPPA, O. J., 164, *169*
KOIZUMI, M., 40, *66*
KORZHINSKII, D. S., 14, 15, *35*

KRACEK, F. L., 40, 45, *66*
KULLERUD, G., 10, 86, 87, *93*
KUME, S., 40, *66*
KURNAKOV, N. S., 8, *12*

LABROW, S., 136, *150*, 161, 168, *169*
LAZARUS, D., 146, *150*
LE COMTE, P., 164, *168*
LEVIN, E. M., 13, 23, 31, *35*, 68
LIANDER, H., 161, *169*
LLOYD, E. C., 161, *169*
LOEWINSON-LESSING, F. Y., 4, *12*
LUNDBLAD, F., 161, *169*
LUTH, W. C., 52, 53, 54, 55, 60, *66*, 99, 100, 108, 112, 117, *118*

MACCHESNEY, J. B., 161, *169*
MACKENZIE, W. S., 1, *12*, 52, 55, *66*, 198, 206, 207, *208*, *209*
MCMURDIE, H. F., 13, *35*
MADORSKY, S. L., 48, *66*
MANNING, W. R. D., 136, *150*, 161, 168, *169*
MAO, H. K., 162, 163, 165, 166, 167, *168*, *169*
MARGRAVE, J. L., 76, *93*
MAYOR, A., 162, *169*
MERWIN, H. E., 6, 7, *12*
MONTGOMERY, P., 154, *169*
MOREY, G. W., 6, 7, *12*, 40, 45, *66*, 94, 95, 97, 117, *118*, 202, *209*
MUAN, A., 9, 170, 184, *190*
MULLHOLLEN, G. L., *118*
MUNOZ, J. L., 172, 182, 183, 185, *190*
MUNRO, D. C., 124, *149*
MYERS, M. B., 154, *169*

NEWBIGGING, D., *150*
NEWHALL, D. H., 122, 128, 136, *150*
NEWTON, M. S., 164, *169*
NEWTON, R. C., 9, *12*, 162, 163, 164, *169*
NOLAN, J., 207, *209*

O'HARA, M. J., 52, 55, 56, 58, 59, 60, 65, *65*, 76, 82, 84, 88, *93*, *93*
ORVILLE, P. M., 202, *209*
ORR, P. L., 164, *169*
OSBORN, E. F., 9, 29, *35*, 170, *190*, 193, *209*

PATERSON, M. S., 151, *169*
PIRSSON, L. V., 5, *12*
PRESNELL, D. C., 46, *66*, 115, *118*

RANKIN, G. A., 6, 67, *93*
RECORD, R. G. H., 89, 91, 92, *93*
REEHER, J. R., 55, *66*
RICCI, J. E., 13, *35*
RINGWOOD, A. E., 162, *169*

Author Index

ROBBINS, C. R., 13, *35*
ROBERTS, H. S., 83, *93*
ROBERTSON, E. C., 122, *149*
ROBERTSON, J. K., 108, *118*
ROBIE, R. A., 206, *209*
ROEDDER, E., 46, 49, 50, *66*, 193, 196, 198, *209*
ROESER, W. F., 70, 75, *93*
ROOSEBOOM, E. H., 164, *168*
ROOYMANS, C. J. M., 99, 105, *118*
ROSENBERG, P. E., 161, *169*
ROY, R., 52, 55, *66*, 152, 153, 154, 155, *169*
ROYSTER, R. H., 48, *66*

SAHA, F., 40, *65*
SATO, M., 89, 90, 91, 92, 93, *93*, 172, 174, 183, *190*
SAUREL, J., 142, *150*
SCHAIRER, J. F., 7, *11*, *12*, 29, 33, *34*, *35*, 41, 43, 44, 45, 48, 49, *65*, *66*, 70, 72, 75, 79, 80, 82, 83, 84, *93*, 126, *150*, 193, 199, 207, *208*, *209*
SCHREIBER, H., *150*
SEIFERT, F., 163, *169*
SHAW, H. R., 60, 64, *66*, 115, *118*, 173, 183, 187, 188, 189, *190*
SHEPHERD, E. S., 6, 7, *12*, 67, *93*
SIMMONS, G., 151, *169*
SKIPPEN, G. B., 171, 172, 177, 180, 181, 182, 183, 185, *189*
SMITH, F. G., 13, 34, *34*, *35*
SMITH, J. V., 162, 164, *169*
SMYTH, F. H., 7, *12*, 119, *150*
SOSMAN, R. B., 68, *93*
SPETZLER, H., 141, *150*
STEWART, D. B., 194, *209*
STROMBERG, H., 154, *169*
SYROMYATNIKOV, F. V., 8, *12*

TAYLOR, N. W., 8, *12*
THOMPSON, J. B., 16, *35*
TILLEY, C. E., 33, *35*
TODD, S. S., 164, *169*
TURNER, F. J., 3, *12*, 13, 22, 29, *34*, *35*, 202, *208*
TURNOCK, A. C., 184, *190*
TUTTLE, O. F., 8, 94, 97, 98, 99, 100, 106, 108, 112, 117, *118*, 206, 207, *209*

VERHOOGEN, J., 3, *12*, 13, 22, 29, *34*, *35*, 202, *208*
VODER, B., 142, *150*, 151, *169*
VON LIMBACH, D., 194, *209*

WALDBAUM, D. R., 206, *209*
WASHINGTON, H. S., 5, *12*
WATT, G., 4, *12*
WEALE, K. E., 124, *150*
WEISBROD, A., 13, *34*
WEISS, J. D., 146, *150*
WENSEL, H. T., 75 93
WILLIAMS, D. W., 112, 113, 114, 116, *118*, 162, 165, *168*, *169*
WILLIAMS, F. J., 8, *12*
WILLIAMSON, E. D., 124, *149*
WONES, D. R., 172, 173, 174, 175, 177, 187, *189*, *190*
WRIGHT, F. E., 6, 67, *93*
WYLLIE, P. J., 8, 9, 10, *12*, 94, 108, *118*, 151, 161, 162, 164, *168*, *169*

YODER, H. S., 8, *12*, 33, *35*, 48, *66*, 87, *93*, 119, 120, 122, 123, 124, 125, 126, 128, 143, 146, *150*

ZUSSMAN, J., 22, *34*, 193, *208*, *209*

Subject Index

ACF diagram, 30
activities of solids and liquids, 171
AKF diagram, 30
Alkemade line, definition of, 25
anvils, *see* tetrahedral anvil
arc welding, 72, 107–108, 184
'basalt' system, 206
belt apparatus, 9, 151, 159–160
    furnace for, 160
    sample assembly for, 159, 160
    temperature–pressure—range of, 160
boron nitride, use in furnace for piston cylinder apparatus, 157
Bridgman seals
    inverted type, 135, 139
    unsupported area type, 95, 124, 126, 128, 130, 134–136, 141
buffers, 87, 97, 171–187
    definition of, 172
    equilibrium in, 186–187
    fluorine buffers, 182, 185
    graphite buffers, 172, 178–180, 182, 183, 184, 186
    oxygen buffers, 172–178, 179, 180, 183, 184, 185, 186, 187
    preparation of, 183
    shorthand notation for, 185–186
    techniques for, 183–186
    theory of, 171–183
calibration of pressure and temperature in solid-media devices, 152, 158, 162–167
    piston-in, piston-out techniques, 163, 164
    pressure calibration, 163–165
    pressure sensitive reactions, 164
    range of pressure corrections, 163
    range of temperature corrections, 166, 167
    temperature calibration, 165–167
calorimetry, in determination of equilibrium, 196, 204, 205, 206
ceramic tubes
    for furnaces, 77, 79, 80
    for thermocouples, 72, 105
closed system, 97
    definition of, 15
cold resistance (R_o), 78

component
    definition of, 14
    fixed, 16
    mobile, 15, 16
computer techniques in predicting phase relations, 13, 34
condensed system, 30
    definition of, 20
congruent melting, definition of, 17
coning and quartering, 63
control of oxygen fugacities by hydrogen–water vapour mixtures, 187–189
correlation of experimental results with natural data, 206–208
critical phenomena
    definition of, 17
    pressure, 17
    temperature, 17
degree of freedom, definition of, 14
disequilibrium, 194, 196
electrolyte, 90, 93
entropy, 2, 20, 22, 23, 202, 205
equilibrium, 2–3, 17, 64, 171, 191, 194
    determination of, 196–200
    divariant (bivariant), 25, 31, 32, 34
    invariant, 15, 20, 21, 25, 31, 32, 33, 34
    relation to grain size, 196, 199
    relation to kinetics, 196, 197, 200–203, 205
    relation to starting materials, 196, 198, 200, 202, 203–204
    trivariant, 25, 34
    univariant, 20, 21, 25, 31, 32, 33, 34
eutectic, 20, 21, 23, 25
    definition of, 19
experiments at atmospheric pressure, 10–11, 67–93
    furnaces for, 76–84
    sample preparation procedures, 85–87
    under inert and controlled atmospheres, 87–93
    with silicates, 85–86
    with sulphides, 86–87
externally heated pressure vessels, 7, 8, 94–116
    Barnes vessel, 115
    closures for, 94, 96, 97, 99, 109, 110, 113, 114, 116, 117

Subject Index 215

externally heated pressure vessels—(cont.)
  furnaces and temperature recording for, 104–106
  hazards and precautions, 109, 115–117
  heat treatment of, 117
  modified Tuttle-type vessels, 99
  molybdenum alloy vessels, 112, 113, 114, 116
  Morey-type vessels, 94–97
  Nimonic 105 vessels, 112, 117
  operation of Tuttle-type vessels, 100–110
  performance of, 110–115
  preparation of samples for, 106–110
  pressure systems and accessories for, 102–104
  Rene 41 vessels, 111, 112, 117
  Shaw-type vessel, 115
  special uses of, 115
  stainless steel vessels, 112
  Stellite 25 vessel, 111, 112, 187
  Tuttle-type vessels, 8, 97–110
  TZM vessels, 111, 113, 114
filler rod (back-up rod), 99, 105, 109, 110, 114, 116, 117
flow diagram, definition of, 33
free energy, 2, 16, 18, 20, 23, 37
  definition of, 22
  determination of equilibrium, 197, 198, 199
  relation to buffers, 175, 176
  relation to kinetics, 200, 201
  relation to starting materials, 64, 203, 204, 205
furnaces, 5, 7, 36, 38, 47, 52, 74, 76–84
  for preparation of iron-bearing glasses, 49, 50
  making, 41, 42, 43, 79
  pot, 41, 42, 43, 48
  quenching, 79–82
  resistance, 38, 47
  thermal-gradient, 82–83
  windings for, 7, 77–79
fugacities
  definition of, 171
  $f(CH_4)$, 180
  $f(CO)$, 178, 179, 180
  $f(CO_2)$, 178, 179, 180
  $f(H_2)$, 172, 173, 175, 177, 178, 180, 181, 182, 184, 185, 187, 189
  $f(H_2O)$, 173, 177, 180, 181, 187, 189
  $f(H_2S)$, 181
  $f(HS)$, 181
  $f(O_2)$, 49, 55, 89, 90, 92, 172, 173, 174, 175, 176, 177, 178, 179, 180, 181, 185, 187, 189
  $f(S_2)$, 181, 182
  $f(SO_2)$, 181
  $f(SO)$, 181

fugacity coefficient, 173, 180, 189
  definition of, 175
gas mixing apparatus, 88, 89
gas welding, 107–108
Geophysical Laboratory Temperature Scale, 68
'granite' system, 206
historical development, 3–10
hot resistance (R_h), 78, 142, 143
identification of phases, 11, 191–196
  composition of phases, determination of, 193
  electron microprobe, 87, 192, 194–195
  fritting, significance of, 192
  optical methods, 75, 85, 87, 148, 192–193, 195
  ratios of phases, determination of, 193, 194
  solid solution, determination of, 192, 193, 194
  textures, use of, 192, 199
  X-ray diffraction methods, 87, 148, 191, 192, 193–194
incongruent melting
  definition of, 17
intensification ratio, definition of, 136, 137
internally heated pressure vessels, 7, 119–149
  Burnham, Holloway, and Davis (1969) vessel, 133–134
  closures for, 121, 127, 128, 130, 131, 134, 135, 148
  furnaces for, 126, 130, 131, 141–145, 147
  Goldsmith and Heard (1961) vessel, 126–129
  Harwood Engineering Co. Inc. vessel, 129–133
  heat treatment of, 120, 124, 126, 134
  insulation in, 121, 125, 126, 131, 143, 144
  operation of, 147–148
  power leads for, 121, 122, 124, 126, 127, 130, 131, 134, 144, 148, 149
  pressure equipment for, 123, 136–141
  pressure medium in, 122, 128
  pressure range of, 119
  quenching of, 148
  safety precautions for, 148–149
  sample holders for, 121, 126, 127, 130, 131, 146–147, 148
  temperature controllers for, 145–146
  temperature equipment for, 141–147
  thermal gradients in, 121, 125, 128, 130, 141, 142, 143, 148
  thermocouple leads for, 121, 124, 125, 127, 130, 131, 145, 148, 149
  Yoder's (1950a) vessel, 8, 122–126
International Temperature Scale, 68

Subject Index

kinetics, 2, 3, 14, 65, 191
  relation to equilibrium, 196, 197, 200–203, 205
lever rule, definition of, 22
limitations of experimental petrology, 1–3
liquid immiscibility, 19, 21, 195
liquidus, definition of, 17
'Ludox', 53, 54, 56, 57, 58
manganin pressure cell, 103, 126, 132, 134
  definition of, 141
mechanical amalgamator, 62
membranes, 115, 184, 185, 186, 187, 188, 189
metastability, 196, 197
  relation to kinetics, 198
  relation to starting materials, 37, 56, 61, 62, 64, 65, 199, 203
modal analysis, 63, 193, 206, 207
multi-anvil device, 161
Newhall packing closure, 127, 128
normative analysis, 206, 207
Ohm's law, 78
open system, 15
opposed anvil device (simple squeezer), 9, 151, 152–155
  pressure determination in, 154
  pumping system for, 152
  range of pressure and temperature, 152, 153
  sample holder for, 153, 154
  sample size in, 154, 155
  thermocouple for, 155
optical pyrometer, 5, 69, 76
Ostwald's law, 201
oxygen sensing cells (oxygen probes), 89–93
partial volatile pressures, control of, 9, 10, 87, 170–189
  $p(CO_2)$, 97, 170, 178, 180
  $p(CO)$, 170, 178, 180
  $p(F)$, 170
  $p(H_2)$, 97, 170, 173, 188, 189
  $p(H_2O)$, 97, 173, 188
  $p(O_2)$, 88, 90, 91, 92, 170, 173, 178, 186
  $p(S)$, 170
peritectic (reaction point), definition of, 21, 22
phase, definition of, 14
phase diagrams, 11, 16–34
phase rules, 2, 13–17, 21, 27
  condensed, 15, 25
  Gibbs, 13, 14–15
  Goldschmidt, 15
  Korzhinskii, 15–16
piercing point, definition of, 32
piston cylinder apparatus, 9, 151, 152, 155–159
  furnace for, 156, 157, 158

piston cylinder apparatus—(cont.)
  insulation in, 156, 157
  power leads for, 156, 157, 158
  pressure determination in, 156
  pressure systems for, 156
  pressure vessel for, 156, 157
  quenching of, 158
  range of pressures and temperatures, 155
  sample assembly for, 157
potentiometer, 69, 70, 71, 73, 106
pressure media, 100, 102, 113, 114
pressure systems and accessories for externally heated vessels, 102–104
  pressure gauge, 103, 109
  pressure reservoir, 103, 110
  pumping system, 103
  tubing and fittings, 103, 104, 117
  valves, 103, 109, 110
pressure systems and accessories for internally heated vessels, 136–147
  high-pressure valves, 119, 131, 139–141, 149
  intensifiers, 102, 119, 122, 131, 134, 136–139, 149
  leak detection, 141
  packings, 137–139, 141
  pressure measurement, 126, 129, 141
  rupture discs, 139, 149
  tubing and fittings, 139, 149
principles of experimental petrology, 1
problems of application of experimental results, 191–209
pyrophyllite, as insulator, 143, 157, 159, 160, 161
quenching method, 6, 11, 47, 195
  for externally heated pressure-vessels, 96, 103, 104, 109, 116
  for internally heated pressure-vessels, 148
  for solid-media apparatus, 158
  one atmosphere experiments, 85
  principle of, 67
quenching rig, 74, 79, 82, 85, 93
  design of, 80, 81
quench products, 192, 194, 195
radioactive tracers, use in gel testing, 60
representation of experimental data, 2, 15, 16, 17
  four-component systems (quaternary), 30, 31, 33
  multi-component systems, 13, 30–34
  one-component systems (unary), 13, 15, 16, 17–20
  three-component systems (ternary), 13, 17, 23–30
  two-component systems (binary), 13, 16, 20–23
reversibility of reactions, 189, 198
  definition of, 197

sample containers, types of,
  gold tubes, 73, 87, 106, 107, 174, 183, 184, 185, 192
  iron crucibles, 49
  platinum foil and tubes, 73, 74, 85, 96, 97, 106, 107, 157, 158, 183, 184, 185, 192
  silica tubes, 48, 86
  silver–palladium tubes, 107, 174, 184, 185
  silver tubes, 106, 183, 184, 185
simplicity of experimental systems, 2, 208
solid-media apparatus, 9, 119, 151–168
  operation, maintenance, safety precautions, 167–168
  pressure and temperature range of, 151
solid solutions in synthetic systems, 19, 20, 21, 22, 25, 26, 27, 30
  determination of, 192, 193, 194
  related to free energy, 23
solubility experiments, 6, 7, 11, 115, 119, 171, 204
solvus, definition of, 21
starting materials, 3, 36–65
  alkali loss from glasses, 37, 45, 47, 48, 51, 64, 65
  choice of, 36–39
  dry and wet mixtures, 39, 61–63, 203, 204
  gels, 37, 39, 40, 51–60, 62, 64, 65, 203, 204
  glasses, 37, 39–51, 61, 62, 64, 65, 203, 204
  homogeneity of, 36, 39, 41, 47, 48, 50, 56, 59, 60, 61, 62, 63, 64, 65, 87

starting materials—(cont.)
  natural starting materials, 36, 39, 63–64
  preparation and testing of, 39–64
  size of, 39, 41, 52, 56, 57, 62, 63, 64
sublimation curve, definition of, 17
surface tension, relation to grain size, 199
temperature calibration, 84–85
temperature controllers, 71, 83–84, 106, 145–146
temperature measurement, 67–76
temperature scales, 67–68
tetraethylorthosilicate (TEOS), 52, 53, 57, 58, 59
tetrahedral anvil, 9, 151, 160–162
  pressure and temperature range of, 160
  sample size, 160
thermocouples, 5, 69–76
  accuracy of, 70
  calibration and construction, 71–76
  lead wires for, 71 73
  reference junction for, 71, 73
  sheathed, 70, 99, 105, 145
  types and temperature ranges, 69–70
thermodynamics, 2, 13, 18, 22, 34, 170, 183, 196, 198, 206
three-phase triangle, definition of, 29
tie line, definition of, 29
triple point, definition of, 17
tungsten carbide pistons, 152, 153, 154, 159, 160
types of experimental petrology, 10–11
vaporous, definition of, 17
variac, 83, 127, 146
voltage stabilizer, 83, 127

MAR 22 1991

QE
433
E33

OCT 23 1974